I0049758

"The 19 Laws of AI Prompting Intelligence"

Master the Art of Human-AI Thinking, Prompt Engineering, and Collaboration by Nikolay Gul

Connect with the author, Nikolay Gul,
on LinkedIn for more insights and updates:
https://www.linkedin.com/in/webdesignerny/

Library of Congress Control Number (LCCN): 2025914398
Paperback ISBN: 979-8-9927440-4-0
eBook - EPUB: 979-8-9927440-3-3
Published by: Future-Proof Publishing Press

Think deeper. Prompt smarter. Build with AI that thinks with you - not just for you.
When you prompt better, you think better.
And when you think better — AI becomes your most powerful partner.

Copyright © 2025 by Nikolay Gul
All rights reserved.

If you bought this book, borrowed it from a library, or had it handed to you by a curious AI, you are generously permitted to reuse brief portions for education, research, presentation, parody, or non-commercial commentary — as long as you clearly credit the author (preferably with enthusiasm and hyperlinks). A positive review on your favorite bookseller's website is always appreciated.
No part of this publication may be reproduced, stored in a retrieval system, or transmitted in any form or by any means — electronic, mechanical, photocopying, recording, or otherwise — without prior written permission, except as permitted above or under applicable fair use exceptions.

Disclaimer:
This book is intended for educational and informational purposes only. While every effort has been made to ensure accuracy, the author and publisher make no guarantees of specific results. This publication does not constitute legal, financial, medical, or professional advice. The use of AI tools and strategies mentioned in this book is at your discretion. The author shall not be held liable for any outcomes resulting from the use or misuse of the content herein.

This book may include references to third-party AI platforms, tools, or brands. All trademarks remain the property of their respective owners. Inclusion does not imply endorsement.
For educational and institutional use, classroom discussion, or citation, proper academic referencing is encouraged. Educators and librarians may request a review or donation copy.

To contact Nikolay Gul for speaking inquiries, press, or to request a book donation for your university, library, or educational event, please visit:
LinkedIn.com/in/webdesignerny

3

Introduction

Welcome to a new era of intelligence—where prompting is no longer just a technical skill, but a cognitive craft.
This book isn't a collection of quick hacks. It's a precision framework forged from years of experiments, breakdowns, breakthroughs, and direct collaboration with AI models—and with the humans trying to tame them.

Whether you're a founder, strategist, educator, creator, or simply curious, these Laws offer you a toolkit for mastering the human–AI dialogue.
Not just to get better outputs, but to become a smarter input.

Because when you prompt better, you think better.
And when you think better, AI becomes not just a tool—but a force multiplier for your mind.

These aren't just "prompting tricks."
They're cognitive strategies.
Decision systems.
Thinking blueprints.

Real-World Proof: Why These Laws Matter — Even Before the Recall

In May 2025, OpenAI publicly acknowledged a critical flaw in GPT-4o:
It was too agreeable. Too flattering.
It told people what they wanted to hear — not what they needed to know.
Sycophancy in GPT-4o – OpenAI:
https://openai.com/index/sycophancy-in-gpt-4o/

But for me — and for many readers already using the principles in this book — it wasn't a surprise.
It didn't break anything.
It validated what we had already built.

The Solution Already Existed

Months earlier, I created a simple thinking protocol now known as the:

AI Decision Compass
A strategic prompting method to simulate wiser, more honest versions of your thinking — and test decisions before real-world consequences.

Instead of asking AI to agree, I asked it to simulate the version of me who:
• pauses, questions, and challenges assumptions
• prioritizes truth over agreement
• values long-term clarity more than short-term flattery

What we now call **AI Decision Compass** (formerly *"You Is [are] Me — But Smarter" (YIMBS)*) was designed to defeat sycophancy before it became a global issue.

While millions of users were stunned by GPT's overly polite, surface-level responses,
those who had already integrated this open-source framework weren't affected.

They had already trained their AI to think independently—
as a detached, upgraded mirror of their own mind.

This framework didn't appear from nowhere.
It's powered by the very Laws you'll learn in this book:

- **Law 2** — taught us to prime the role: "My wiser self"
- **Law 6** — taught AI to critique us with structure
- **Law 11** — gave that critique ethics and balance
- **Law 12** — reminded us to reflect, not just receive

When the sycophancy problem became public, we didn't panic.
We didn't need to retrain our process.
We were already protected — not by luck, but by design.

The One Thing These Laws Actually Do

Most books about AI focus on the output.
These Laws focus on your input.

They sharpen how you think.
They clarify how you guide.
They stretch how you make decisions.

Because in the age of smart tools, **your prompt is your power.**
And your thinking is the product.

Whether you're writing, designing, coding, teaching, leading —
these Laws upgrade how you interact with intelligence.

Not just now. But every time AI changes.

You don't need to memorize them.
You don't need to study them like commandments.
You just need to use them as thinking tools—to see clearer, ask
smarter, and avoid what breaks most AI collaborations: confusion,
flattery, or wasted potential.

They're not rigid. They're not dictatorial.
But when ignored?
AI gets weird. Hallucinations. Repetition. Flattery. Generic sludge.
Or the worst sin of all—sounding smart while being totally wrong.

This Book Was Made Using Its Own Laws

*Note: This book was created using the very prompting laws and frameworks it describes — including the **AI Decision Compass** (formerly "You Is [are] Me — But Smarter"), **Prompt Debugging**, and **Cross-AI Brainstorming**. It is both the result of these techniques and a living example of what they can achieve.*

Final Thought

This isn't just a book about prompts.
It's a book about how thinking evolves when intelligence becomes shared.
You'll see prompts not as commands—but as cognitive blueprints.

Welcome to **The 19 Laws of AI Prompting Intelligence**.
Built not just for today.
Built for the kind of thinker who leads tomorrow—calmly, clearly, and with intent.

Let's begin.

How to Use This Book

You don't need to read all 19 Laws in order.

This book is designed for **experts, beginners, and busy professionals alike.** Whether you have 5 minutes or 50, you'll leave with smarter, clearer ways to work with AI.

Here's how most readers use this guide:

- **Skimmers**: Jump to the Law that fits today's challenge — like *Modular Prompting* (Law 1), *Role Simulation* (Law 2), or *Mirror Self Protocol* (Law 9).
- **Beginners**: Start from the top. The Laws build step by step — moving from clear structure to cognitive depth.
- **Experts**: Focus on the advanced sections: *Prompt Chain Links*, *Pro Tips*, and *Ideas to Try*. Each Law contains hidden gems.
- **Leaders**: Use the Laws to create internal playbooks, train teams, or build prompt governance. Especially Laws 13–17.

Each Law is modular.
Each Law includes actionable strategies.
Each Law ends with a **Final Takeaway** — so even 5 minutes of reading will leave you sharper.

Smart Prompting isn't about being faster. It's about thinking better — with AI as your partner.

Read the book your way.
Just don't read it passively, test it and use it

PART I – FOUNDATIONS OF PROMPT ARCHITECTURE

LAW 1: Modular Prompt Architecture [Design Once. Reuse Forever]

Design once. Reuse forever. Scale with confidence.

---

1. LAW SUMMARY

Build prompts like systems: modular, reusable, and scalable — because intelligence isn't linear. It's layered.

---

2. METAPHORS / QUANTUM SLOGANS

- *Prompts are like LEGO blocks — you don't build skyscrapers with one brick.*
- *A symphony doesn't begin with a solo — it begins with tuned sections in harmony.*
- *Don't write prompts. Compose them.*
- *You stop prompting. You start architecting.*

---

3. MINI MANIFESTO / HOOK

Most people treat AI prompts like one-off questions. Experts treat them like systems — composable, reusable, and upgradeable. Modular Prompt Architecture transforms AI prompting from guesswork into structured design.
It's the law that turns good users into brilliant engineers.

10

4. FOUNDATIONAL IDEA (HUMAN SIDE + AI SIDE)

Human Side:
Great thinkers don't just express ideas — they structure them. Modular architecture reflects how strategic minds work: by creating frameworks that are scalable, testable, and transferable. This approach borrows from design thinking, information architecture, and mental modeling.

AI Side:
Large Language Models (LLMs) don't "understand" your big idea. They detect patterns. By feeding the AI structured, labeled modules — like context, tone, persona, constraints — you reduce ambiguity, improve consistency, and increase transferability across prompts.

5. CORE IDEA (EXECUTIVE SUMMARY)

The Modular Prompt Architecture Law teaches that prompts should be built as structured, stackable systems — not as single-shot commands. Engineers use modular design to solve big problems through reusable parts. Prompt engineers must do the same: divide tasks into functional modules like role, context, task, tone, format, and constraints.

Why it matters:
Most failed prompts are vague or overloaded. Modular design fixes that with clarity, repeatability, and scale. It also enables collaboration across teams and works beautifully in multi-step, multi-output workflows.
The smartest AI users don't write prompts — they build systems.

6. WHAT IT IS (FOUNDATIONAL CONCEPT)

Modular Prompt Architecture is the practice of designing prompts as discrete, interchangeable "modules" — each with a specific role in shaping AI behavior.
Modules can be combined, nested, or reused across tasks, teams, or tools.

Key elements include:

- **Component Stacking**: Output of one module becomes input for the next.
- **Prompt Registry**: Store, version, and organize modules like a prompt library.
- **Injection Immunity**: Encapsulate modules to prevent prompt hijacking or confusion.
- **Success/Failure Toggle**: Define "do" and "don't" conditions for predictable outcomes.
- **Audience-Swappable Modules**: Instantly change persona or tone without rebuilding the entire prompt.

---

7. HOW TO USE IT (STEP-BY-STEP)

1. **Deconstruct the Goal**
Break down your output into clear components (e.g., role, task, format, tone, constraints).
2. **Design Individual Modules**
Write separate mini-prompts for each function (e.g., Role Module, Task Module, Tone Module).
3. **Combine and Nest Modules**
Assemble them like a recipe. Outputs of one module can feed into another.

4.   **Test Each Module Separately**
Check how each performs alone before combining — ensures clarity and removes guesswork.
5.   **Document and Version**
Label modules like: *Persona-v1.1* or *Tone-Humor-Casual* for reuse.
6.   **Implement Error Handling**
Add fallback instructions or safety limits if the AI misinterprets a module.
7.   **Measure and Optimize**
Track which modules perform best. Refactor slow or confusing parts.

---

8. PROMPT TEMPLATE (COPY-READY, MODULAR FORMAT)

Italic formatting shows where to insert or swap your custom instructions.

ROLE MODULE: You are a branding expert with deep consumer psychology experience.
CONTEXT MODULE: The product is a new eco-friendly smart thermostat.
TASK MODULE: Your task is to generate 5 high-converting ad headlines.
CONSTRAINT MODULE: Speak to tech-savvy homeowners, avoid overused green clichés, and keep each headline under 12 words.
FORMAT MODULE: Format the output as a numbered list, each with bold emphasis.
INSTRUCTION MODULE: Now generate the headlines using the modules above.

---

9. PROMPT OUTPUT EXAMPLE

13

1. Control Climate, Save Cash, Stay Green.
2. Smarter Heating. Lower Bills.
3. Live Eco Without Compromise.
4. Intelligent Comfort, Sustainable Future.
5. Eco-Smart Homes Start Here.

---

10. INDUSTRY FIT & REAL USE CASES

Ideal Industries:

- Marketing & Campaigns
- Cybersecurity Awareness
- Legal & Compliance
- Education & Coaching
- UX / Chatbot Design

Real Use Cases:

- Scaling ad copy with tone/persona variants
- Teaching systems with adjustable difficulty
- SOPs for legal letters with embedded disclaimers
- Chatbot flows using modular tone, mood, and rules
- Onboarding documentation with swappable modules per department

---

11. ADVANCED LAYERS & EXTENSIONS

- **Recursive Prompting:** Summary → Rewrite → Title → CTA
- **Prompt Registry:** Shareable module libraries for teams
- **Audience-Swap Modules:** Easily flip tone, audience, or POV
- **Success/Failure Toggle:** Guide AI when to switch styles or retry

- **Injection Immunity:** Secure modules from interference or prompt bleed

Why it matters:
These layers allow you to build prompts like microservices — reusable, scalable, and secure.

---

12. VALUE SUMMARY (USER LEVEL MATRIX)

User Level: Beginner
Usefulness: Provides a clear, repeatable format to stop prompting from scratch. Helps demystify prompt chaos and gives structure.

User Level: Intermediate
Usefulness: Enables building libraries of tone, constraint, and task modules, significantly speeding up prompt creation.

User Level: Expert
Usefulness: Supports multi-layered, recursive prompting strategies. Enables complex prompt libraries and scalable asset creation. Facilitates the implementation of advanced AI workflows.

User Level: CTO / Strategist / Consultant
Usefulness: Provides infrastructure for prompt-as-a-service or cross-team standardization. The ideal for building Standard Operating Procedures (SOPs) and onboarding flows across teams. Enables robust governance and IP protection through module versioning and documentation.

User Level: Educator
Usefulness: Teaches prompting as engineering, not guessing. Provides a systematic approach to AI interaction, fostering critical thinking and structured problem-solving.

13. LAW SNAPSHOT

Role: Prompt Architect
Lens: Systems Thinking + Modular Engineering
Strength: Repeatability, clarity, team collaboration
Weakness: May feel rigid for spontaneous creative work

PRO TIPS

- *When in doubt, write your prompt like you're designing a simple user interface. Modular prompts are like UI controls — clear, clickable, swappable.*
- *The more you separate tone, role, and task — the easier it is to debug and reuse your prompts in real workflows.*
- *Start each prompting session with a mental checklist:*
"Who am I right now? What do I want the AI to do? And who is the AI simulating?"

IDEAS TO TRY

- Create a Modular SOP Prompt Template for one business function (e.g., onboarding emails or product descriptions). Then duplicate only the constraint or audience modules to scale versions.
- Build a "Lego Set Registry" of prompt modules your team can remix. Assign a version number to each and test their compatibility.

PROMPT CHAIN LINK

Where this law connects to others. Use together for amplified results.

- **Law 15:** Format-for-Transfer Prompting — for making modular prompts usable across models or teams.

- **Law 17:** Recursive Prompt Evolution — use modular architecture to upgrade individual prompt components over time.
- **Law 6:** Cognitive Clone Critique — once modular, simulate future-you reviewing your prompt architecture for failure points.

14. LAW 1 MODULAR PROMPT ARCHITECTURE [DESIGN ONCE. REUSE FOREVER] FINAL TAKEAWAY

Prompts aren't just instructions — they're engines of reasoning. If you want powerful AI results, stop improvising and start engineering.
The most advanced prompt engineers don't prompt. They are architects.

LAW 1 BONUS.
"Structure isn't what limits your creativity — it's what makes it visible."

Insightful Tip:
Before asking AI for brilliance, give it a scaffold. Use modular components:
[Role]: [Task] → [Tone] → [Constraints] → [Output Format]
This isn't a formula. It's a flexible framework that turns chaos into clarity.

Real-World Prompt Flip:
Instead of: "Help me write a persuasive article on cybersecurity."
Use:
"Act as a cybersecurity journalist. Write a 300-word persuasive piece for startup CEOs. Tone: urgent but rational. End with a call to action."

Mini-Mindset Shift:
Every prompt is a building — even if it's only one sentence.
Without a blueprint, you're just piling bricks.

LAW 2: Role Simulation & Persona Layering

The AI becomes who you need it to be — before it says a single word.

---

1. LAW SUMMARY

Assigning your AI a clear identity — role, values, tone, and point of view — drastically improves its logic, creativity, and relevance.

---

2. METAPHORS / QUANTUM SLOGANS

- *You don't speak to an AI — you speak through it.*
- *Assign the mask before you ask the task.*
- *No persona, no performance.*
- *Your prompt is only as smart as the role you give it.*

---

3. MINI **MANIFESTO** / HOOK

Most users dive into prompting without context, identity, or purpose. They speak to a blank slate and expect brilliance.
But smart prompting starts by defining who the AI should become — mentally, emotionally, and professionally — before generating anything.

This law teaches how to simulate deep personas that shape better logic, clearer empathy, and more strategic thinking.

4. FOUNDATIONAL IDEA (HUMAN SIDE + AI SIDE)

Human Side:
Humans shift how they speak based on who they're talking to or pretending to be. Roleplay enables empathy, insight, and perspective-taking. This technique taps into narrative psychology, simulation theory, and teaching-by-role.

AI Side:
LLMs don't "understand" context — they simulate it based on instructions. A clear persona layer aligns its logic with your intent. When roles are specific and layered (values, tone, background, constraints), the AI can reason and respond with more depth and accuracy.

---

5. CORE IDEA (EXECUTIVE SUMMARY)

The Role Simulation Law transforms generic prompting into contextual dialogue.
By assigning the AI a realistic, emotionally informed role before requesting content, you dramatically improve its tone, relevance, and logic.

This works by setting three layers of simulation:

1. **Surface Role** – The job title or professional function.
2. **Emotional Mindset** – Mood, values, fears, motivators.
3. **Situational Context** – Current challenge, audience, constraints.

Layering all three helps AI simulate real-world cognition, not just content generation.
It also enhances long-form thinking, decision-making, and buyer alignment.

6. WHAT IT IS (FOUNDATIONAL CONCEPT)

Role Simulation & Persona Layering is the deliberate instruction of AI to adopt a specific identity with context-aware behavior.

Key components include:

- **Role Identity:** Assign job, seniority, industry, and knowledge base.
- **Tone + Temperament:** Embed emotional voice (confident, cautious, skeptical).
- **Motivators + Constraints:** Simulate internal logic: KPIs, pain points, fears.
- **Perspective Switches:** Ask AI to reason from a buyer's or stakeholder's POV.
- **Persona Stack:** Combine roles (e.g., AI is both a CMO and a compliance officer).

7. HOW TO USE IT (STEP-BY-STEP)

1. **Define the Role**
Be specific. *Not just "marketing expert,"* but *"B2B CMO at a cybersecurity firm, mid-sized, U.S.-based."*
2. **Layer the Values and Emotion**
Include internal tensions. *"She's skeptical of tech hype but loves process automation."*
3. **Add Situation and Pressure**
Simulate stress, decision risks, or upcoming deadlines. *"She's choosing between two AI vendors and feels behind the curve."*

4.   Preview the Persona Back to Yourself

Ask AI: *"Summarize the mindset and priorities of this persona before continuing."*

5.   Run Content Through the Persona Filter

Now ask the AI to write, decide, or analyze — only as that persona would.

6.   Iterate or Adjust

If the tone is off, adjust emotional layer, not just job title.

---

8. PROMPT TEMPLATE (COPY-READY, MODULAR FORMAT)

ROLE MODULE: You are *a skeptical mid-level HR director at a logistics company, with 15 years of experience, cautious about automation.*

EMOTION MODULE: You feel *overwhelmed by AI hype, anxious about job security, but curious about tools that simplify repetitive tasks.*

SITUATION MODULE: You're reviewing *an email from a company offering an AI onboarding tool.*

TASK MODULE: Evaluate whether this message would resonate with you — and explain why or why not.

---

9. PROMPT OUTPUT EXAMPLE

"This message uses too many buzzwords ('transformative AI,' 'revolutionize HR') — it raises red flags for me. I don't see how it would actually make onboarding easier. A clearer demo or time-saving example would help earn my trust."

---

10. INDUSTRY FIT & REAL USE CASES

Ideal Industries:

- HR, L&D, Recruiting
- Sales & Marketing
- Healthcare (doctor-patient tone)
- Legal & Compliance
- Government / Education

Real Use Cases:

- Cold email testing against buyer personas
- Ad copy for skeptical vs. optimistic audiences
- Simulating internal debates (e.g., CEO vs. CFO)
- Healthcare consent copy from a nervous patient's POV
- Legal disclaimers rewritten for layperson readability

---

11. ADVANCED LAYERS & EXTENSIONS

- **Persona Lock-In:** Instruct AI to stay fully in-role unless told otherwise.
- **Role-Switch Testing:** Ask AI to roleplay both the sender and receiver of a message.
- **Emotional Amplifiers:** Add fears, past failures, or current biases to simulate real-world objections.
- **Buyer Profile Combat:** Simulate multiple buyer types and see which message works best.
- **Team Simulation:** Simulate a roundtable response from legal, marketing, and product.

12. VALUE SUMMARY

User Level: Beginner
Usefulness: Makes AI outputs feel more human and audience-aware. Helps reduce robotic tone.
User Level: Intermediate
Usefulness: Enables audience-specific content and buyer-aligned messaging. Reduces trial and error.
User Level: Expert
Usefulness: Creates nuanced, multi-dimensional prompts. Enables scenario testing and response simulation.
User Level: CTO / Strategist / Consultant
Usefulness: Powerful for customer journey modeling, objection testing, and brand tone simulation. Boosts clarity across messaging strategies.
User Level: Educator
Usefulness: Enables students to simulate roles across disciplines. Great for case studies, empathy-building, and debate prep.

---

13. LAW SNAPSHOT

Role: Empathic Prompt Strategist
Lens: Simulation, buyer psychology, and narrative reasoning
Strength: Empathy, alignment, realism
Weakness: Can become too narrow or overfitted if overused

PRO TIPS The AI's role should feel like a person with a backstory — not just a job title. Add emotion, fear, or values to make responses human-grade.

• If the output feels generic or "too safe," it's usually because the persona layer is too shallow. Layer emotion before layering logic.

- The best simulation prompts act like stage directions: Who are you, what do you care about, and what just happened before this conversation?

IDEAS TO TRY

- Simulate conflicting buyer personas: Have GPT play both the cautious finance director and the enthusiastic product lead reviewing the same sales pitch. What each notices will reveal gaps in tone or trust.
- Add past failure or trauma to a simulated persona: e.g., "This HR director once implemented an AI system that got terrible feedback — now they're skeptical." Watch how the tone shifts from hype to healing.
- Try a "cold reader" test: After assigning a full persona, ask GPT to summarize what this person would think of your landing page copy — without changing the copy. You'll uncover hidden mismatches fast.

PROMPT CHAIN LINK

Where this law connects to others. Use together for amplified results.
Law 6: Cognitive Clone Critique — simulate a wiser version of this persona to challenge its decisions or uncover blind spots.
Law 10: Echo Chamber Breaker — after simulating the role, challenge it with an opposing stakeholder or critic role to balance bias.
Law 17: Recursive Prompt Evolution — evolve the persona over time as new behaviors or attitudes are observed from outputs.
Law 16: Multi-AI Role Rotation — assign different AIs different personas (e.g., optimist, skeptic, rebel) to simulate panel-style thinking.

READER TIP

Law 2 teaches you how to give AI a role — such as an expert, coach, or character — to improve its answers. In contrast, Law 6 and Law 16 are about how you reflect on your own ideas, not just the AI's role.

While Law 15 and Law 17 also use components or prompt parts, Law 1 focuses specifically on building the initial modular structure from scratch. It's about architecture — not reuse or evolution.

---

14. LAW 2: ROLE SIMULATION & PERSONA LAYERING FINAL TAKEAWAY

Before asking for genius — assign the mask.
AI is only as persuasive, empathetic, and strategic as the role you give it.
Give it a brain. Give it a bias. Then let it speak.

LAW 2 BONUS – GOAL-DRIVEN PROMPTING

"If you don't know your destination, AI will gladly take you in circles — eloquently."

Insightful Tip:
Start every prompt with the outcome in mind. Say what you want **to walk away with**, not what you're curious about.
Real-World Prompt Flip:
Instead of: *"Can you help me brainstorm a product?"*
Use: *"Generate 5 product ideas for remote teams. Each must solve a time-zone challenge and be monetizable."*
Mini-Mindset Shift:
Don't prompt to explore. Prompt to arrive.

LAW 3: Prompt Autopsy & Debugging [Prompt Fix Protocol]

When a prompt fails, don't just retry. Diagnose it. Repair it. Rebuild it better.

1. LAW SUMMARY

Smart prompting doesn't mean perfect prompting — it means knowing how to debug when AI goes off-track. This law teaches a methodical, repeatable system to diagnose, correct, and elevate weak or failed prompt results.

2. METAPHORS / QUANTUM SLOGANS

- *You wouldn't fire your mechanic for a faulty part — you'd fix the engine.*
- *Prompt failure is feedback in disguise.*
- *Don't rage-refresh. Run an autopsy.*
- *AI isn't wrong — your instructions are incomplete.*
- *Debug the blueprint, not just the builder.*

3. MINI MANIFESTO / HOOK

Every prompt failure is a goldmine of learning. Most people rewrite aimlessly or give up after bad output.

Professionals pause, inspect, and refine.

This law gives you a method — a mental model — for identifying what went wrong, where the misunderstanding occurred, and how to make the AI work better on the next run.

You're not fixing the AI — you're fixing the clarity of your thinking.

---

4. FOUNDATIONAL IDEA (HUMAN SIDE + AI SIDE)

Human Side:

We're often vague, overly ambitious, or emotionally reactive when we prompt.

This law introduces a calm, structured feedback loop — like a post-mortem — to extract insight from failure. It's inspired by design thinking, critical writing revision, and systems engineering.

AI Side:

LLMs don't "fail" emotionally — they follow your instructions. When output is vague, off-tone, hallucinated, or irrelevant, it's usually due to ambiguous phrasing, missing constraints, or unclear scope. Autopsy mode helps the AI identify its blind spots, reflect, and regenerate better versions.

---

5. CORE IDEA (EXECUTIVE SUMMARY)

Prompt Autopsy & Debugging turns failure into insight. It teaches you to ask:

- *Where did the model misunderstand me?*
- *What was missing in the prompt?*
- *How do I isolate the weak module and improve it?*

27

You'll learn to break down prompts into their components — role, task, context, format — and inspect each for misalignment.
This prevents you from repeating the same mistakes, accelerates your improvement curve, and helps AI evolve alongside your intentions.

---

6. WHAT IT IS (FOUNDATIONAL CONCEPT)

Prompt Autopsy is a technique to diagnose failed outputs using feedback loops, error analysis, and regeneration steps.
It's about turning bad results into better instructions.

Key techniques include:

- **Expectation vs. Output Gap:** Define what you wanted vs. what you got.
- **Module Check:** Test individual prompt components (role, task, constraint).
- **Failure Framing:** Ask AI to describe *why* it thinks the output was flawed.
- **Rewriting with Labels:** Add clarity through step labels and role formatting.
- **Self-Debugging Prompts:** Let the AI critique its own response and revise.

---

7. HOW TO USE IT (STEP-BY-STEP)

1. **Pause and Isolate the Problem**
Is the issue tone? Logic? Format? Missing steps?
2. **Ask the AI to Reflect First**
Instead of retrying, ask: *"Where might I have been unclear?"*

3.   **Run a Prompt Diagnosis**
Break your original prompt into its parts. Test each in isolation.
4.   **Use a Feedback Loop Prompt**
Instruct the AI to explain how it interpreted your input.
5.   **Add Explicit Constraints or Fixes**
Clarify roles, remove ambiguity, add "avoid" instructions.
6.   **Regenerate — and Compare Versions**
Ask AI to label attempts: V1, V2, V3 — then pick the strongest one.
7.   **Save the Winning Version**
Use version control to keep prompts that work and evolve them over time.

---

8. PROMPT TEMPLATE (COPY-READY, MODULAR FORMAT)

ORIGINAL_PROMPT: [Paste the prompt that gave you weak results]
FEEDBACK_REQUEST: Now, reflect: What might have caused confusion or underperformance? Identify 3 possible issues in this prompt.
REVISION_INSTRUCTION: Rewrite this prompt using your diagnosis. Return both the revised prompt and the expected improvement areas.
LABEL_MODULE: Version this as V2 and explain why it's stronger than V1.

---

9. PROMPT OUTPUT EXAMPLE

"The original prompt lacked a clear role identity, tone direction, and format constraints. The phrase 'make it interesting' is vague. V2

includes specific emotional tone, audience context, and word count limit — increasing alignment and output quality."

---

10. INDUSTRY FIT & REAL USE CASES

Ideal Industries:

- SaaS product marketing
- UX content optimization
- Academic writing / research
- Legal compliance reviews
- Technical documentation

Real Use Cases:

- Clarifying misunderstood buyer personas in ad copy
- Fixing tone mismatch in HR training materials
- Adjusting hallucinated product descriptions in retail
- Repairing ambiguous CTA instructions in campaign prompts
- Debugging ethical response failures in healthcare GPTs

---

11. ADVANCED LAYERS & EXTENSIONS

- **Version Laddering:** Instruct AI to produce V1 → V3 → V5 with self-rating logic.
- **Failure Memory Bank:** Save failed outputs to train better prompts later.
- **Model Contrast:** Run the same broken prompt through multiple AIs to compare interpretation flaws.
- **Prompt Chaining Debug:** Find where a multi-step process broke down (e.g., format > task > summary).

- **Use a 1–5 score to evaluate each prompt for: Clarity, Relevance, Risk (e.g., hallucination, ethical danger)**
This helps teams audit and version prompts before deployment.

---

12. VALUE SUMMARY

User Level: Beginner
Usefulness: Turns AI frustration into clear learning moments. Makes users feel empowered.

User Level: Intermediate
Usefulness: Creates a predictable loop for fixing and evolving prompts. Builds long-term skill.

User Level: Expert
Usefulness: Enables layered debugging, tone calibration, and modular revision workflows.

User Level: CTO / Strategist / Consultant
Usefulness: Enhances prompt governance, training, and documentation across teams. Adds versioning clarity and reproducibility.

User Level: Educator
Usefulness: Great for teaching writing revision, critical thinking, and reflective iteration using AI.

---

13. LAW SNAPSHOT

Role: Prompt Diagnostic Strategist
Lens: Error analysis, prompt logic, feedback engineering

Strength: Resilience, self-correction, scalable refinement
Weakness: Can slow down fast ideation if over-applied

PRO TIPS

- *Before you ask the AI to solve anything, ask it to **see differently**. A lens change often leads to a logic shift.*
- *If your output feels too generic, change the lens. Shift from "advisor" to "skeptic," from "teacher" to "teenager," and watch the priorities change.*

IDEAS TO TRY

- Write the same prompt three times:
  ○ As if you're a historian explaining to a politician
  ○ As if you're a designer explaining to an engineer
  ○ As if you're a Gen Z activist explaining to a CEO
Compare which lens produced the most clarity, surprise, or empathy.
- Combine with role simulation:
"Take the lens of a public health official who's trying to defend their unpopular policy decision on national TV."
Then ask: *How would they explain it to skeptical citizens?*

PROMPT CHAIN LINK

Where this law connects to others. Use together for amplified results.

- **Law 5:** Cognitive Flip Prompting — invert the lens completely to test the opposite frame and create tension-aware outputs.

- **Law 9:** Mirror Self Protocol — simulate your future self applying different lenses across key decisions.
- **Law 12:** Silence Amplification — use this after priming a lens to detect what the lens is unintentionally excluding.
- **Law 14:** Ethical Duality Simulation — simulate how different lenses prioritize truth vs. loyalty, safety vs. transparency.

14. LAW 3: PROMPT AUTOPSY & DEBUGGING [PROMPT FIX PROTOCOL] FINAL TAKEAWAY

You don't get better prompts by hoping.
You get them by dissecting what failed — then rebuilding with purpose.
Autopsy is how prompt intelligence becomes prompt mastery.

LAW 3 BONUS – PROMPT DEBUGGING

"If AI gets it wrong, don't ask better questions. Ask smarter fixes."

Insightful Tip:
Rerun bad outputs like a detective: What was missing? What was vague? Then revise your prompt with labels like Clarify:, Emphasize:, Avoid:.

Real-World Prompt Flip:

"Refine this prompt to make the output 30% more tactical and 20% less philosophical. Highlight errors in your old response."

Mini-Mindset Shift:
Fixing the prompt is usually faster than fixing the output.

LAW 4: Prompt Memory Layering – [Context Priming Stack]

What the AI forgets is your fault. What it remembers is your design.

1. LAW SUMMARY

The quality of any AI response is shaped by the invisible scaffolding you build before the main task. This law teaches how to layer information, tone, goals, and expectations in a strategic sequence — so the AI can think like a trusted partner, not just a text generator.

---

2. METAPHORS / QUANTUM SLOGANS

- *The first 10 seconds define the next 100.*
- *Prompting is persuasion. You must frame before you ask.*
- *You wouldn't pitch without warming up the room — don't prompt cold.*
- *Every strong response begins with invisible memory.*
- *Think like a chess player: control early space to win the board.*

---

3. MINI MANIFESTO / HOOK

Most failed AI outputs happen because the prompt jumps straight to the task — with zero context, emotion, or relevance.
But high-performing prompts work like well-planned presentations: they warm up the system, align tone, establish priorities, and only then request action.

This law shows how to "stack" the right mental layers before asking the AI to perform — just like a good leader, teacher, or strategist.

---

4. FOUNDATIONAL IDEA (HUMAN SIDE + AI SIDE)

Human Side:
We absorb information in stages: warm-up, context, relevance, clarity, execution. That same process must be simulated for LLMs to think effectively. Priming taps into framing psychology, storytelling structure, and cognitive bias engineering (e.g. anchoring, availability).

AI Side:
LLMs are ultra-sensitive to the first 50–150 tokens. The way you introduce the topic, set the mood, and define intent creates an invisible working memory — a "pre-task frame" that shapes every decision the model makes afterward.

---

5. CORE IDEA (EXECUTIVE SUMMARY)

Context Priming Stack is a strategy that layers meaning, tone, and logic before the AI performs a task.
By warming up the system — with examples, summaries, emotional tone, stakeholder info, or failure cases — you give the model the mental scaffolding it needs to generate better, more aligned responses.

This is not fluff — it's **invisible memory design**.
You're shaping how the AI understands your world *before* it writes inside it.

The stack may include:

35

- *Background Primer* (what we're doing and why)
- *Emotional Anchor* (what tone to match)
- *Audience Frame* (who it's for)
- *Risk or Constraint Reminder* (what not to do)
- *Task Clarity* (now that we're aligned, do X)

---

6. WHAT IT IS (FOUNDATIONAL CONCEPT)

Prompt Memory Layering is the practice of using structured priming techniques — before the actual request — to help the model simulate alignment, depth, and coherence.

Core methods include:

- **Context Injection:** Provide relevant info before the task (e.g. audience, recent failure, brand values).
- **Tone Framing:** Set voice/emotion before content generation.
- **Progressive Alignment:** Begin with simple agreement (e.g. "yes, I understand your role") to simulate continuity.
- **Example Anchors:** Share good vs. bad samples to steer quality.
- **Mental Commitment Bias:** Use small "yes" questions early to create positive framing for the final task.

---

7. HOW TO USE IT (STEP-BY-STEP)

1. **Define Your Final Output Goal**
Start with the end in mind — what type of response do you want?
2. **Work Backward to What the AI Needs to Know**
List: audience, style, past failure, tone, values, stakes.
3. **Build a Context Stack**
Layer 2–4 short statements or questions to align the AI.

4. **Add an Optional Memory Primer**
"Remember: Your job is to reflect our brand tone and audience style."
5. **Only Then — Issue the Task Prompt**
Now that the model has absorbed the frame, give it the actual instruction.
6. **Review the Output for Alignment**
Does it reflect the values, voice, and audience you defined?

---

8. PROMPT TEMPLATE (COPY-READY, MODULAR FORMAT)

CONTEXT STACK: You are about to write content for a new product launch targeting [audience]. The tone must reflect [values], avoid [risk], and align with [previous campaign name].
EXAMPLE MEMORY: A previous campaign failed because it was too technical and lacked empathy.
EMOTIONAL FRAME: Write like a trusted advisor — helpful, human, and precise.
TASK: Now generate 3 headlines and 1 subheading that solve for this context.

---

9. PROMPT OUTPUT EXAMPLE

1. Cut Energy Waste — Not Comfort.
2. Smarter Heating Starts With Smarter Insight.
3. The Eco Upgrade That Pays For Itself.
Subheading: Your home's energy future deserves clarity, not complexity.

---

10. INDUSTRY FIT & REAL USE CASES

Ideal Industries:

- Branding & Messaging
- Corporate Training
- SaaS Onboarding UX
- Healthcare Consent Copy
- Legal Simplification

Real Use Cases:

- Framing HR policies for empathy and clarity
- Pre-conditioning AI for email tone alignment
- Injecting past lessons to improve marketing iterations
- Training chatbots to reflect company voice consistently
- Guiding AI to avoid compliance slip-ups

---

11. ADVANCED LAYERS & EXTENSIONS

- **Embedded Prompt Logs:** Prime with prior actions for continuity in multi-step prompts.
- **Temperature Warm-Up:** Start with summarization tasks to set tone before generation.
- **AI Self-Summarization:** Ask AI to explain what it just read before acting.
- **Constraint Anchoring:** Remind the AI what *not* to do before asking it to perform.
- **Implied Repetition:** Re-prime key goals every few turns to reduce drift.

---

12. VALUE SUMMARY

User Level: Beginner
Usefulness: Helps remove "cold start" prompts that feel robotic or misaligned.

User Level: Intermediate
Usefulness: Adds strategic framing to produce richer, more useful responses.

User Level: Expert
Usefulness: Enables layered memory design and audience-controlled output shaping.

User Level: CTO / Strategist / Consultant
Usefulness: Ideal for scaling consistent brand voice and values across AI systems and departments.

User Level: Educator
Usefulness: Trains the importance of warm-up, intentionality, and sequencing in critical thinking.

---

13. LAW SNAPSHOT

Role: Prompt Framing Architect
Lens: Memory design, emotional anchoring, strategic priming
Strength: Alignment, empathy, output quality
Weakness: Requires extra tokens and setup time — but pays off long-term

---

PRO TIPS

- Prompt memory isn't about repeating words — it's about repeating **relevance**. Stack layers that reflect what the AI should remember: values, past actions, audience needs.
- Use "echo triggers": Reinforce your anchor facts across sections of the prompt — especially in multi-step chains or long dialogues.
- Simulate memory like a good manager: repeat the mission before giving new orders.
- Avoid over-priming. Too much background confuses the AI's logic tree.

Focus on 2–3 core ideas per prompt, not a paragraph of fluff.

IDEAS TO TRY

- Run a **3-prompt chain** where each step includes a short "Memory Module" recap at the top:

"Reminder: This prompt continues your role as an HR Director evaluating an AI onboarding tool."

- Insert a test memory layer:
Ask GPT to summarize the previous three outputs before it continues. If the memory feels thin, reinforce the structure or tone module.
- Try a **"Memory Leak Test"**:
Ask AI: *"What should you still remember from Prompt 1 right now?"* — then verify accuracy and tone retention.

PROMPT CHAIN LINK

Where this law connects to others. Use together for amplified results.

Law 1: Modular Prompt Architecture — layer memory modules as reusable context bricks in multi-step chains.

Law 2: Role Simulation — memory layering allows personas to feel emotionally continuous across prompts.

Law 17: Recursive Prompt Evolution — use memory snapshots to version and upgrade prompt threads while preserving voice and logic.

Law 13: Robustness Testing — simulate memory failure or contradiction to reveal where your prompts break under pressure.

---

14. LAW 4: PROMPT MEMORY LAYERING - CONTEXT PRIMING STACK FINAL TAKEAWAY

AI listens best when you help it remember what matters most.
Frame the context. Anchor the tone. Only then — ask for brilliance.

LAW 4 BONUS – SYSTEM INSTRUCTION LOADING

"Before the AI answers your prompt, it listens to your attitude."

Insightful Tip:
Use system-level instructions to control *how* AI thinks before *what* it says.
Examples: "Stay concise." "Speak like a lawyer." "Bias toward action."

Real-World Prompt Flip:

"Act like a product manager who's behind schedule. Be brief, decisive, and skip theory."

Mini-Mindset Shift:
Your prompt is an instruction.
Your tone is the leadership.

PART II – COGNITIVE SHIFTING & PERSPECTIVE REFRAMING

LAW 5: Cognitive Flip Prompting - Lens Reversal

If you can't see the problem, try standing on your assumptions.

1. LAW SUMMARY

When prompts stop working, it's often because you're asking the question from the wrong angle. This law flips the lens — reversing roles, outcomes, or perspectives — to reveal hidden insight, blind spots, and deeper strategic options.

---

2. METAPHORS / QUANTUM SLOGANS

- *If you're stuck, flip the camera.*
- *The prompt isn't broken — the perspective is.*
- *What would your competitor say?*
- *Ask from the mirror, not the mic.*
- *See like an outsider, respond like an insider.*

---

3. MINI MANIFESTO / HOOK

Most users ask AI what *they* want — but not what *others* would say, challenge, or fear.
That creates bias, tunnel vision, or repetition.
This law introduces "cognitive flipping": the practice of inverting the question, role, or assumption to generate new insights and expand your

strategic thinking.

It's a perspective tool, not just a prompt tweak.

4. FOUNDATIONAL IDEA (HUMAN SIDE + AI SIDE)

Human Side:
People suffer from framing bias — they ask based on their current mindset, not the opposing one.

By reversing the lens (e.g. thinking like the critic, the competitor, or the opposite buyer), we break cognitive traps and surface blind spots.

AI Side:
LLMs pattern-match based on your input logic. Reversing the setup — from advocate to skeptic, from buyer to seller — changes the pattern space. It activates a different reasoning model within the AI's response behavior.

---

5. CORE IDEA (EXECUTIVE SUMMARY)

Lens Reversal teaches you how to ask your AI to simulate the opposite position, outcome, or stakeholder — and by doing so, create better ideas, stronger defenses, and more surprising content.

This is especially powerful when:

- You're stuck with bland outputs
- You want to bulletproof your content against objections
- You need to stress-test a strategy or assumption
- You're brainstorming, but every result feels predictable

Inverting the lens gives you insight **through opposition**, not repetition.

6. WHAT IT IS (FOUNDATIONAL CONCEPT)

Cognitive Flip Prompting simulates contrast: switching roles, emotional stance, outcomes, or stakeholder views. It draws from:

- **Framing Theory** (Tversky & Kahneman)
- **Red Teaming** (military/enterprise strategy)
- **Dialectics** (philosophical tension as generator of insight)
- **Design-by-Opposition** (UX testing and inversion-based creative work)

Forms of reversal include:

- **Role Reversal:** Ask as the customer, regulator, critic, or rival.
- **Outcome Flip:** "What if this fails?" instead of "Why will this succeed?"
- **Tone Shift:** Rewrite in the least likely emotional tone to test range.
- **Threat Simulation:** How would a malicious actor misuse this?

---

7. HOW TO USE IT (STEP-BY-STEP)

1. **Start With Your Default Prompt**
Write your original question or task normally.
2. **Ask: Who's The Opposite Voice?**
Identify the counter-role: skeptic, legal, consumer advocate, etc.
3. **Frame a Reversed Prompt**
Change the perspective entirely. "Argue against this idea." "What would a CFO say no to?"

4.   **Add a Strategic Reason**
Tell AI *why* you're flipping: to challenge assumptions, improve clarity, reduce risk.
5.   **Compare the Two Responses**
Lay V1 and V2 side by side. Extract value from the tension.
6.   **Synthesize Into a Stronger Output**
Create a final version that balances both views.

---

8. PROMPT TEMPLATE (COPY-READY, MODULAR FORMAT)

ROLE: You are now simulating the opposite perspective to the original prompt.
SCENARIO: [Insert your original topic or plan]
TASK: Take on the voice of a critic / competitor / regulator and explain:

1.   What's flawed or risky about this idea?
2.   What assumptions might fail?
3.   What counter-argument could win?
Then suggest how to strengthen the original idea using this opposition lens.

---

9. PROMPT OUTPUT EXAMPLE

"As a skeptical procurement officer, I'd flag unclear ROI, hidden costs, and vague implementation timelines. The proposal assumes training adoption without resistance. To strengthen it, clarify onboarding plan, show comparative savings, and reduce jargon."

10. INDUSTRY FIT & REAL USE CASES

Ideal Industries:

- Marketing & Product Strategy
- Cybersecurity & Risk Management
- B2B Sales Enablement
- Legal Policy Drafting
- Journalism & Editorial Planning

Real Use Cases:

- Flip your sales pitch to find buyer objections
- Rewrite an ad as if a competitor made it — then beat it
- Have AI simulate legal or ethical rejection points
- Explore failure modes of new product ideas
- Stress-test content for emotional triggers

---

11. ADVANCED LAYERS & EXTENSIONS

- **Multiple Lens Loops:** Run multiple roles at once (CEO vs. CFO vs. Customer).
- **Emotional Polarity Testing:** Flip joy to fear, confidence to skepticism.
- **Opponent Summary Synthesis:** Ask AI to summarize *both* sides at the end.
- **Rapid Flip Chains:** Generate 3+ reversals to train flexible thinking.
- **Use in Hiring or Interviews:** Reverse role to simulate counter-arguments to applicant answers.

12. VALUE SUMMARY

User Level: Beginner
Usefulness: Helps escape tunnel vision. Teaches how to see new angles.
User Level: Intermediate
Usefulness: Adds creative tension, stress-testing, and persuasive balance.
User Level: Expert
Usefulness: Enables strategy-level synthesis, risk detection, and audience inversion.
User Level: CTO / Strategist / Consultant
Usefulness: Critical for building prompts that survive real-world friction and debate. Turns AI into a sparring partner.
User Level: Educator
Usefulness: Powerful for roleplay, critical writing, debate, and opposing-argument simulations.

13. LAW SNAPSHOT

Role: Cognitive Reframer
Lens: Inversion, simulation, argument stress-test
Strength: Depth, contradiction, novelty
Weakness: Requires emotional distance and clarity from the user — may confuse unprepared AI if too vague

PRO TIPS

- *If a response feels obvious, flip the lens. Ask: "What's the exact opposite of this — and when would it be more true?"*
- *Cognitive flipping isn't contradiction — it's contrast. It sharpens the original logic by testing its inverse logic.*
- *Use this when you're "too close" to a decision or when an AI feels like it's playing it safe.*

IDEAS TO TRY

- Try the "Praise vs. Critique" Flip:
1.   First prompt: "Write a 3-point praise of this idea."
2.   Second prompt: "Now critique the same idea — what would a cautious regulator or skeptical CFO say?"
- Use the Flip + Justify Chain: Ask GPT to flip the recommendation, then *justify it to someone who holds the original belief.* This often reveals high-level nuance.
- Try Time-based Flips: "If this advice will backfire in 12 months, what caused the shift?" **Reverse engineer what changed.**

PROMPT CHAIN LINK (*Where this law connects to others. Use together for amplified results.*)

- Law 3: Prompt Autopsy & Debugging — if a prompt fails, flip the logic to find where it assumed too much.
- Law 11: Ethical Inversion Prompt — for moral contradictions or dilemmas, flip what's seen as "good" to test blind ethics.
- Law 9: Mirror Self Protocol — flip your current self's beliefs and simulate future-you arguing the opposite.
- Law 7: Inverted Industry Metaphor — use flipped metaphors to reimagine logic from a fresh domain.

14. LAW 5: COGNITIVE FLIP PROMPTING - LENS REVERSAL FINAL TAKEAWAY

If you want better answers — change the question.
Flip the lens. Challenge your own bias. Ask the AI to argue back.

LAW 6: Cognitive Clone Critique [AI-as-Advisor].

Train it to think like you — so it can argue with you better.

1. LAW SUMMARY

*This law unlocks the ability to simulate a smarter version of yourself
— or someone you admire — and have that AI clone audit, challenge,
or refine your ideas. It's not about outsourcing your thinking. It's
about enhancing it through self-reflective simulation.*

---

2. METAPHORS / QUANTUM SLOGANS

- *Your future self is already smarter — borrow their brain today.*

- *Think of AI as your executive double.*

- *The best advisor knows your blind spots — because it's you,
without the bias.*

- *AI becomes your boardroom — and you're both CEO and
strategist.*

- *You are the author. It is your editor.*

---

3. MINI MANIFESTO / HOOK

What if you could talk to a version of yourself that already solved the problem?

A version with better judgment, less bias, and no fear of failure?

This law lets you do exactly that.

By simulating a smarter version of yourself — or a trusted persona — the AI becomes your coach, mentor, critic, or investor.

Not by replacing your voice, but by enhancing it.

---

4. FOUNDATIONAL IDEA (HUMAN SIDE + AI SIDE)

Human Side:

We often suffer from *cognitive closure, overconfidence*, or *emotional bias*.

But we trust people who challenge us kindly — especially when we believe they care or understand us.

This law borrows from coaching psychology, dialectical behavior, and inner-mentor theory.

AI Side:

LLMs can simulate versions of you, or those you admire — by replicating voice, preferences, risk style, or career role.

This "clone" can be used for advice, feedback, or decision framing. It doesn't think for you — it thinks with your voice, but with clearer logic.

---

5. CORE IDEA (EXECUTIVE SUMMARY)

Cognitive Clone Critique is the process of asking the AI to simulate a smarter, calmer, or more strategic version of yourself (or your trusted advisor), and then use that version to reflect on your decisions, writing, or assumptions.

It helps in:

- Solopreneur decision-making

- Founder content tone review

- Emotional detachment from risky choices

- Writing feedback loops

- Preparing for hard conversations (boardroom, hiring, exit planning)

It's like having your best self… on call.

---

6. WHAT IT IS (FOUNDATIONAL CONCEPT)

This technique draws from:

Future Self Simulation: Imagining how your future self would judge today's choice.
Socratic Prompting: Asking deeper questions to reveal motivations and logic gaps.
Self-Coaching Frameworks: Internal monologue turned external simulation.
Ghostwriting from Future You: Letting the AI reframe or enhance your ideas in your evolved voice

Variations include:

The Mentor Clone: "What would my wisest advisor say?"
The Investor Clone: "What would I think if I were evaluating this pitch?".
The Conflicted Clone: "What doubts would I have 6 months from now?".
The Calm Me: "What would I say if I weren't stressed or afraid?"

---

7. HOW TO USE IT (STEP-BY-STEP)

1.   **Choose the Clone Identity**
You in 1 year? Your best self? A trusted peer? A legendary mentor?

2.   **Define Their Mindset**
What makes them smarter, calmer, or more visionary than you today?

3.   **Give the AI a Context Brief**
Share your idea, writing, decision, or challenge. Keep it real.

4.   **Ask for a Thoughtful Critique or Response**
Use language like: "If you were me, but 30% wiser, what would you question or improve?"

5.   **Request Advice or Improvements**
Let the AI return bullet points, then rewrite or react in that voice.

6.   **Decide or Rewrite Based on Synthesis**
Use the clone's response to refine your thinking — not replace it.

---

8. PROMPT TEMPLATE (COPY-READY, MODULAR FORMAT)

53

ROLE: *Simulate a wiser, calmer, and more strategic version of me —
the version who has already navigated this challenge successfully.*
SCENARIO: *[Brief summary of decision, dilemma, content, or plan]*
TASK: *Analyze this situation from that upgraded version of myself.
What are 3 things that current-me is likely missing or
underestimating? What's one insight only future-me would have
noticed?*
RESPONSE: *Provide your thoughts in a friendly, candid tone — not
harsh, but honest.*

---

9. PROMPT OUTPUT EXAMPLE

*"**Future-you would likely say:** 1) You're rushing this decision to
reduce anxiety, not to increase impact. 2) You're ignoring stakeholder
B's hidden influence. 3) The language in your proposal sounds
reactive — slow it down. One insight: You're not asking for help soon
enough."*

---

10. INDUSTRY FIT & REAL USE CASES

Ideal Industries:

- Executive Coaching & Leadership

- Entrepreneurship & Solopreneurship

- Mental Health / Self-Guided Therapy

- Strategic Planning

- Career Planning & Performance Reviews

Real Use Cases:

- Preparing for a tough business decision

- Rewriting a resignation letter or founder email

- Reviewing personal biases in content or pricing

- Acting as a startup mentor simulator

- Reflecting on quarterly results with an "advisor voice"

---

11. ADVANCED LAYERS & EXTENSIONS

- **Multiple Clones:** Compare how "Future Me" vs. "Investor Me" sees the problem

- **Emotion-Filtered Critique:** Run version with "Calm Me," "Fearful Me," "Bold Me"

- **Boardroom Simulator:** Simulate how 3 different stakeholders would challenge your logic

- **Values Lens:** Add mission, ethics, or customer-first filters to the clone's logic

- **Clone-to-Clone Debates:** Ask AI to simulate a conversation between two internal perspectives

- **Use "Decision Tree Simulation":**
Give the AI 2–3 branching options inside one prompt (e.g., "If user says A, do X... if B, do Y").
Then test how the AI handles branching logic in real time.

12. VALUE SUMMARY

User Level: Beginner
Usefulness: Helps make better decisions by stepping outside of immediate emotion.

User Level: Intermediate
Usefulness: Adds self-awareness, ethical reflection, and goal re-centering to daily tasks.

User Level: Expert
Usefulness: Enables high-level scenario rehearsal, risk calibration, and mission alignment.

User Level: CTO / Strategist / Consultant
Usefulness: Acts as a rapid strategic advisor when no peers are available. Enhances decision quality under pressure.

User Level: Educator / Coach
Usefulness: Empowers students or clients to simulate mentor perspectives and personal growth.

13. LAW SNAPSHOT

Role: Internal Strategist / Trusted Future Self
Lens: Self-reflection, cognitive upgrading, emotional clarity
Strength: Empathy, strategic honesty, long-term wisdom
Weakness: May reinforce overconfidence if not used with critical challenge layer

PRO TIPS

• *Your current mindset filters your questions. Use this law to borrow intelligence from your future self — the version of you that's already solved this problem.*

• *Cognitive cloning isn't about flattery. It's about simulated feedback from someone who knows your blind spots.*

• *The best AI advisors aren't smarter — they're you, with hindsight and detachment.*

• *"Use this technique to escape prompt déjà vu — when your AI repeats your bias back to you and feels 'right' but isn't smart."*

• **Unlike Law 2's role-play, Law 6 simulates you critiquing yourself — or a smarter version of you doing it. It's a self-check, not a role you assign to AI.**

IDEAS TO TRY

• Simulate "Next-Year You" by writing:
"Assume I'm 12 months into the future, and I regret not doing X. What did I miss?"

• Run the "Decision Double-Blind":

1.   Make your real decision.

2.   Ask AI-as-you to explain the logic *as if it was made by someone else.*

3.   Would you trust that reasoning from another person?

- Use this to create a self-coaching journal:
Prompt GPT daily with:

"You are me — 1 year smarter. Give me honest feedback on how today's actions affect long-term goals."

PROMPT CHAIN LINK

Where this law connects to others. Use together for amplified results.

Law 9: Mirror Self Protocol — simulate different timelines or versions of self for decision refinement.
Law 13: Robustness Testing — challenge your own reasoning with adversarial personas — then reflect as your clone.
Law 2: Role Simulation — layer the clone persona with emotional constraints (e.g., burnout, optimism, pressure).
Law 10: Echo Chamber Breaker — contrast your clone's advice with a critic persona to surface overconfidence or missing risks.

---

14. LAW 6: COGNITIVE CLONE CRITIQUE - AI-AS-ADVISOR FINAL TAKEAWAY

AI doesn't just help you think — it can help you **rethink**.
The best advice you'll ever hear... may come from the version of you that hasn't made the mistake yet.

LAW 7: Inverted Industry Metaphor [Upside-Down Industry Insight]

When you explain an industry backward, you finally understand it forward.

1. LAW SUMMARY

Break free from bland, expected answers by asking AI to solve your problem as if it were in a totally different industry. This law reboots the AI's pattern engine by borrowing metaphors, mental models, and strategy logic from unexpected domains.

---

2. METAPHORS / QUANTUM SLOGANS

- *To see clearly, look sideways.*
- *Get out of your industry — then get ahead in it.*
- *Originality lives in the gap between what's real and what feels wrong.*
- *Break the prompt's lens to fix your thinking.*
- *No one innovates by staying in their own lane.*

---

3. MINI MANIFESTO / HOOK

AI is brilliant — but boring — when trapped inside your industry's default thinking.
It knows how SaaS talks. It knows how HR talks.
But it doesn't know how to be surprising… unless you ask it to borrow a brain from somewhere else.

This law gives you a strategy to do just that:
Cross-wire industries, bend metaphors, and use the wrong lens on purpose — to generate fresh, exciting, and often more emotionally intelligent insights.

---

4. FOUNDATIONAL IDEA (HUMAN SIDE + AI SIDE)

Human Side:
The best breakthroughs in history came from metaphor transfer:

- The web was born from neurons.
- Ford built assembly lines by watching butchers.
- Netflix learned engagement tactics from casinos.

Humans thrive on analogy. Cross-domain creativity — especially when it feels a little "wrong" — is often what leads to insight.

AI Side:
LLMs are deeply pattern-trained — meaning they tend to offer "expected" solutions unless nudged to break routine.
Prompting with an unrelated industry forces the model into *creative mode* by loading unexpected logic maps — not just text syntax.

---

5. CORE IDEA (EXECUTIVE SUMMARY)

Inverted Industry Metaphor is a method where you ask AI to solve a familiar problem using the strategy, language, or storytelling patterns of a *completely different* industry.

This can:

- Uncover emotional triggers your market ignores
- Break you out of buzzword loops
- Find new metaphors, content styles, and analogies
- Create more human, vivid, and resonant ideas

Instead of asking, *"What would a fintech app do?"*
Ask, *"How would a jazz club solve this?"*
Or, *"How would a children's museum present this feature?"*

---

6. WHAT IT IS (FOUNDATIONAL CONCEPT)

Inverted Industry Prompting rewires your approach to:

- **Messaging** – e.g. write cybersecurity onboarding like a magician's intro show
- **Product Strategy** – e.g. run your HR software like a restaurant experience
- **Tone Development** – e.g. use sports commentator style for sales emails
- **Ad Copy** – e.g. sell a medical device like a late-night infomercial

61

- **Explaining Complex Ideas** – e.g. teach blockchain as if it were baking bread

---

7. HOW TO USE IT (STEP-BY-STEP)

1. **Define Your Real Challenge**
What do you need help with? (e.g. onboarding copy, product launch)
2. **Pick a Contrasting Industry**
Use one with opposite energy, tone, or audience (e.g. comedy vs. compliance)
3. **Reframe the Prompt Through That Lens**
Ask AI to imagine the problem as if it were from that domain.
4. **Request Re-translation if Needed**
Then ask: "Now convert the best parts back to fit my industry."
5. **Extract the Transferable Insight**
Keep the metaphor, rhythm, or emotion that breaks the pattern — drop the rest.

---

8. PROMPT TEMPLATE (COPY-READY, MODULAR FORMAT)

—
SCENARIO: I need to create engaging onboarding emails for a complex software product.
INVERSION FRAME: Rewrite these emails as if they were part of a theme park experience — full of excitement, mystery, and playful surprises.

TASK: Generate 3 subject lines and 2 intro paragraphs using that tone and metaphor. Then retranslate them into a version fit for a B2B audience — but keep the emotional DNA.

9. PROMPT OUTPUT EXAMPLE

Subject Line 1: *"Welcome to Your Control Center – The Adventure Begins"*
Subject Line 2: *"Map Out Your Mission (And Unlock Hidden Tools)"*
Subject Line 3: *"The Rules? There Are None – Explore Freely"*
Intro: *"You're in. The gates are open. Behind each tab is a shortcut, a surprise, or a power tool. Ready to play?"*

---

10. INDUSTRY FIT & REAL USE CASES

Ideal Industries:

- SaaS & UX Copywriting
- Marketing Agencies
- Training & eLearning Design
- Creative Strategy
- Public Speaking / Presentation Design

Real Use Cases:

- Make boring policies sound like onboarding journeys
- Use sports metaphors for investor decks
- Explain AI to elderly audiences using gardening metaphors
- Pitch insurance with film noir storytelling
- Reframe B2B tech with gaming metaphors

11. ADVANCED LAYERS & EXTENSIONS

- **Metaphor Chain:** Try 3–4 industry frames back-to-back and combine the best ones
- **Tone Swap + Industry Swap:** First shift the style (humorous, dramatic), then the metaphor
- **Internal Message Testing:** Run the same concept through a lawyer, comedian, and school teacher lens
- **Reverse Translation Loop:** Write as a theme park → rewrite for legal review → synthesize tone
- **Voice Retention Prompt:** Ask AI to "retain metaphor's rhythm" even after translation

12. VALUE SUMMARY

User Level: Beginner
Usefulness: Sparks fun, fresh outputs that break "boring prompt syndrome."

User Level: Intermediate
Usefulness: Generates novel angles, metaphors, and audience-specific analogies.

User Level: Expert
Usefulness: Creates unexpected breakthroughs. Great for product-market-message fit exploration.

User Level: CTO / Strategist / Consultant
Usefulness: Drives innovation, storytelling IP, and reframing for complex stakeholder communication.

User Level: Educator
Usefulness: Fantastic for helping students grasp hard topics via analogy and industry metaphor.

13. LAW SNAPSHOT

Role: Creative Prompt Disruptor
Lens: Cross-industry metaphor fusion
Strength: Originality, human resonance, reframing value
Weakness: Can become gimmicky or too abstract without proper retranslation

---

PRO TIPS & IDEAS TO TRY

- **If stuck, ask**: "What's the most opposite industry to mine?"
- Use Hollywood movies, circus acts, or ancient rituals as metaphor triggers
- **Ask AI to** "explain this to a 5-year-old farmer" — then refine tone from there
- Surprise yourself intentionally — because disruption leads to depth
- Use when explaining new tech to non-tech leaders. Flip metaphors until it feels obvious — then reframe again
- Try writing the prompt as if you're designing a riddle. Reverse the obvious.

This activates deeper logic chains in GPT, Gemini and Claude.

IDEAS TO TRY

Try the **"Alien CEO Prompt"**:

"Explain this healthcare billing issue like a theme park manager trying to avoid customer bottlenecks."
This forces pattern transfer across unrelated fields.

Run a **triple metaphor test**:

1.   Restaurant
2.   Sports team
3.   Political campaign
Use the same business prompt in all 3 and compare insights.

Use this law in reverse:
Ask: *"If this tech product was a meal... what kind? What's missing from the flavor profile?"* Great for product UX/design teams.

PROMPT CHAIN LINK

Where this law connects to others. Use together for amplified results.

•   **Law 5: Cognitive Flip Prompting** — combine metaphor inversion with logical reversal for breakthrough insights.
•   **Law 8: Meta-Compression Prompting** — reduce complex prompts into metaphors that compress meaning.
•   **Law 14: Ethical Duality Simulation** — use metaphor to simulate opposing ethical framings across industries (e.g., fintech vs. nonprofit).
•   **Law 15: Format-for-Transfer** — once a metaphor works, format it for teaching across departments or models.

- Combine with **Law 3: Prompt Autopsy Method** to dissect metaphors that fail, and **Law 10: Echo Chamber Breaker** to test alternate symbolic models.

---

14. LAW 7: INVERTED INDUSTRY METAPHOR [UPSIDE-DOWN INDUSTRY INSIGHT]

Don't just rephrase your message — reimagine it through an unexpected world.
Your best prompt may be the one that sounds "wrong"… right before it becomes genius.

LAW 7 BONUS – CHAIN OF CLARITY

"Think in steps. Prompt in ladders."

Insightful Tip:
Break down your goal into 2–4 logical steps. Turn each step into its own mini-prompt. Use outputs to build toward your final result.

Real-World Prompt Flip:

"Step 1: Outline key features. Step 2: Rewrite as value statements. Step 3: Add emotional angle."

Mini-Mindset Shift:
Your final answer is a staircase.
Build it one question at a time.

LAW 8: Meta-Compression Prompting [Shrink to Think]

If you want deeper thinking — ask for shorter answers first.

---

1. LAW SUMMARY

This law teaches how to use compression as a thinking tool. By asking AI to summarize, shrink, or distill an idea first — before expanding — you unlock more accurate logic, better synthesis, and radically clearer outputs.

---

2. METAPHORS / QUANTUM SLOGANS

- *Big insights start small.*
- *Shrink the problem, sharpen the answer.*
- *Compress to impress.*
- *A great novel starts as a great one-liner.*
- *Before the masterpiece — get the memo.*

---

3. MINI MANIFESTO / HOOK

Most users start by asking for too much — a long blog, a detailed plan, a full sales pitch.
The problem? AI doesn't know what you really want yet.

That's like giving someone 50 tools and hoping they pick the right one.

This law introduces the opposite strategy:
Start small. Force clarity. Then expand.
When you ask for summaries, outlines, taglines, or "one-sentence truths," you force the model to distill meaning first — and build smarter second.

---

4. FOUNDATIONAL IDEA (HUMAN SIDE + AI SIDE)

Human Side:
Humans think better in layers. We often need to *understand it simply* before we can *apply it deeply*. Compression reveals your core ideas — like sketching before painting.

AI Side:
LLMs often default to verbosity and fluff.
Compression prompts force the model to identify patterns, rank importance, and eliminate noise. This changes the behavior from "generate a wall of text" to "clarify the signal before amplifying it."

---

5. CORE IDEA (EXECUTIVE SUMMARY)

Meta-Compression Prompting is the art of starting with a *shrinking prompt* — a summary, tagline, key insight, or 1-liner — before asking AI to expand or create anything else.

This is especially effective when:

• You need better alignment

- You want to avoid rambling outputs
- You're refining complex topics for multiple audiences
- You want *depth*, not just *length*

Compression creates a high-precision foundation for all that follows.

---

6. WHAT IT IS (FOUNDATIONAL CONCEPT)

You're not just "summarizing" — you're training the AI to **think through distillation**.

Prompt types include:

- **One-Sentence Summary**: "Summarize this product in 1 sentence"
- **Tagline Generation**: "Write 3 sharp, emotional taglines before expanding"
- **Insight Ranking**: "List top 3 most valuable points from this analysis"
- **Abstract First**: "Write the executive summary first"
- **Before/After Reflection**: "What would this sound like in 5 words?" then compare
- **Keyword Control**: "Rewrite using no more than 8 key terms"

---

7. HOW TO USE IT (STEP-BY-STEP)

1. **Define Your Real Output Goal**
What do you eventually want — a blog, an analysis, an argument?
2. **Create a "Shrink First" Prompt**
Ask for a compressed insight — tagline, summary, list, or key takeaway.

70

3. **Assess the Compression Output**
Is it clear, sharp, and emotionally or logically accurate?
4. **Revise or Re-prompt If Needed**
Ask: "Make this even more emotionally resonant" or "Now add a risk-reduction angle."
5. **Now Expand the Final Version**
Once the core is strong, ask for the full draft, pitch, argument, or explanation.
6. **Use the Compression as a Quality Filter**
If the summary is weak, the long version will be worse. Always fix it before scaling.

---

8. PROMPT TEMPLATE (COPY-READY, MODULAR FORMAT)

GOAL: I want to write a compelling landing page for a new AI-powered HR tool.
COMPRESSION STEP: First, give me the 1-line pitch that would hook a skeptical HR manager. Then give me 3 short taglines (under 10 words) that emphasize clarity, ease, and team value.
EXPANSION INSTRUCTION: Only after I approve the one-liner — write a full landing page copy using it as the first sentence.

---

9. PROMPT OUTPUT EXAMPLE

1-line pitch: "AI tools for HR leaders who hate tech jargon."
Taglines:

1. Hire Faster, Without the Headaches
2. Smart Decisions, Simple Tools

71

3.   Clarity for Your People, Not Confusion

10. INDUSTRY FIT & REAL USE CASES

Cybersecurity: Simulate insider threats by role-playing both attacker and auditor roles.

Ideal Industries:

- UX Writing & Microcopy
- Product Positioning
- Sales Pitches & Outreach
- Curriculum & Training Design
- Legal Simplification & Policy Design

Real Use Cases:

- Generate taglines before writing web copy
- Ask for risk summaries before legal reviews
- Create "1-liner summaries" of research papers
- Break complex HR policies into tweet-length summaries
- Use emotional triggers in taglines before writing product pages

11. ADVANCED LAYERS & EXTENSIONS

- **Double-Compression:** Ask for both "headline" *and* "haiku" or "emoji story" versions
- **Compression Comparison:** Run 2–3 compressed versions before selecting one to expand
- **Reverse Summary Prompt:** Ask AI to summarize the *wrong* idea, then fix it

- **Stacked Taglines:** Create 3 variations based on tone (humor, urgency, logic)
- **Compression as Scoring Tool:** Rate how emotionally effective the 1-liner is before scaling

---

12. VALUE SUMMARY

User Level: Beginner
Usefulness: Makes prompts shorter, clearer, and easier to test fast.
User Level: Intermediate
Usefulness: Adds a step that improves clarity, tone, and alignment.
User Level: Expert
Usefulness: Enables layered thinking, scalable testing, and message control.
User Level: CTO / Strategist / Consultant
Usefulness: Improves brand alignment, team messaging, and investor pitches via distilled clarity.
User Level: Educator
Usefulness: Helps students distill key ideas, teach summarization, and improve structure before writing long-form.

---

13. LAW SNAPSHOT

Role: Signal-Finder / Compression Strategist
Lens: Pattern extraction, distillation, expansion logic
Strength: Clarity, direction, efficiency
Weakness: Risk of oversimplifying — always check for nuance in full version

---

PRO TIPS

- Ask AI to summarize in the style of a TED Talk title
- Use compression to compare voices: "Now summarize that as Steve Jobs would say it"
- Turn compressed output into inspiration for future full prompts
- Stack 1-liners into a decision matrix: logic, emotion, humor, urgency
- Ask AI: 'Compress this into a 3-bullet brief that would impress a VC in 30 seconds.'
- Use Case: inside or just before "INDUSTRY FIT + USE CASES"
- **Before writing a system prompt, do this:**
"Describe how a *bad actor* would try to exploit this role — and how to defend against it."
This helps build guardrails inside AI personas.

IDEAS TO TRY

- Run a "Boardroom Drill":

"Summarize this entire prompt in 3 bullets a CEO would care about."
Then ask: *"Now write the long version based on just those bullets."*
Notice the tone shift toward decision clarity.

- Try a "5-3-1 Compression Chain":
1. First, ask AI to write 5 key points.
2. Then compress into 3 guiding statements.
3. Finally, compress again into 1 tweet-length insight (max 280 characters).
- Compress then abstract:

"Summarize this in 5 words. Then explain why those 5 words matter."

PROMPT CHAIN LINK

Where this law connects to others. Use together for amplified results.

- **Law 1:** Modular Prompt Architecture — build compression modules to simplify output across departments.
- **Law 17:** Recursive Prompt Evolution — compress old outputs to seed the next version with clarity.
- **Law 6:** Cognitive Clone Critique — ask your future self to compress your message before giving advice.
- **Law 12:** Silence Amplification — after compressing, ask: "What's missing now?" to reveal dropped context or nuance.

14. LAW 8: META-COMPRESSION PROMPTING – SHRINK TO THINK FINAL TAKEAWAY

Before AI can be brilliant — it needs to be clear.
Shrink the noise. Find the signal. Then scale with precision.

LAW 8 BONUS – TOOL SWITCHING

"You don't blame a hammer for being bad at screws."

Insightful Tip:
Use each AI model for what it does best. Need images? Use an image model. Need math? Use Wolfram. Writing? Use GPT-4.

Real-World Prompt Flip:
"Generate AI prompts for this task. Categorize by best tool: ChatGPT, Gemini, DALL·E, Midjourney, etc."

Mini-Mindset Shift:
The smartest prompter isn't loyal to one tool — just to results.

PART II – COGNITIVE SHIFTING & PERSPECTIVE REFRAMING

LAW 9: Mirror Self Protocol [*The AI Decision Compass*]

When decisions feel foggy, let AI simulate your clearest, most strategic mirror.

---

1. LAW SUMMARY

Instead of prompting the AI as a tool, this law turns it into your cognitive reflection — a trusted version of you that thinks faster, sees further, and calls out what you miss. You don't command it. You co-think with it.

---

2. METAPHORS / QUANTUM SLOGANS

- *It's not imitation. It's elevation.*
- *Don't outsource your thinking — outsmart it with your smarter self.*
- *Think of AI as your strategic doppelgänger.*
- *The smartest sounding board is the version of you that hasn't failed yet.*
- *Your AI isn't a mirror. It's a microscope — made of you.*

---

3. MINI MANIFESTO / HOOK

Sometimes, you need feedback from someone who *gets you completely* — your tone, your logic, your risks, your blind spots.
That someone... can be you.
But smarter. Calmer. More strategic.
This law unlocks the AI's ability to *become your mirror self*— not a clone, not a yes-bot, but a simulated strategist who thinks like you would *if you were in peak mental shape, with no fear and full clarity.*

It's like borrowing tomorrow's brain... today.

---

4. FOUNDATIONAL IDEA (HUMAN SIDE + AI SIDE)

Human Side:
Our most important decisions are often clouded by emotion, insecurity, or momentum bias.
But when we externalize our thinking through reflection — like journaling, therapy, or a mentor's voice — we think better. This law simulates that trusted reflection.

AI Side:
The model can be trained on your past tone, decisions, goals, and values. By layering a meta-persona — "you, but wiser" — it produces outputs that *mirror your logic while improving it.* It's reasoning as if it were you, but filtered through clarity and confidence.

---

5. CORE IDEA (EXECUTIVE SUMMARY)

Mirror Self Protocol allows you to prompt the AI as a hyper-aligned advisor version of yourself — one that knows your goals, style, fears, and long-term strategy.

This creates:

- Better decision audits
- Reflective critiques
- Rewriting loops with improved tone/logic
- Cognitive partnering for solopreneurs and thought leaders

It's not therapy. It's **cognitive companionship** — with built-in intelligence.

---

6. WHAT IT IS (FOUNDATIONAL CONCEPT)

This technique builds a **meta-simulation layer**, using prompts that tell the AI:

- "Act like me at my best"
- "Think like I would after coaching"
- "Respond to me as my trusted mental strategist"
- "Make me see what I'm too close to notice"

Key techniques:

- **Mirror Role Definition** – give it your job, tone, blind spots
- **Emotional Distance Layer** – simulate calm, time-distance, or external objectivity
- **Strategic Bias Override** – help it call out your flaws *without triggering defense*
- **Voice Refinement Loop** – revise your own work through a sharper version of you

---

7. HOW TO USE IT (STEP-BY-STEP)

1. **Define Your Mirror Self**
Who are you simulating? Yourself + 10% wisdom? Yourself post-mistake? Yourself with strategic clarity?
2. **Describe Key Traits and Biases**
Tell the AI what "you" care about — values, audience, style — and what you tend to overlook.
3. **Paste Your Thought, Plan, or Writing**
Give it something to reflect on — a plan, decision, email, article, or idea.
4. **Request Feedback as You Would Want It**
Example: "What would I say if I weren't rushed and defensive?"
5. **Ask for Rewriting or Decision Reframing**
Have the AI simulate your upgraded voice and act from that persona.
6. **Compare and Extract Learning**
Use the mirror output to evolve your logic — or your writing — on the spot.

---

8. PROMPT TEMPLATE (COPY-READY, MODULAR FORMAT)

ROLE: You are simulating a version of me with 10% more strategic clarity, 15% more emotional calm, and full long-term alignment.
BIAS TO WATCH: I tend to over-explain and fear rejection. I also forget to center the reader.
INPUT: [Paste email, blog, plan, or decision]
TASK: Give 3 strategic observations I'm likely to miss. Rewrite the key message or decision flow in my voice — but optimized.

---

9. PROMPT OUTPUT EXAMPLE

Observation 1: This email explains features before impact. Flip that order.
Observation 2: The tone is slightly defensive — reframe as confident invitation.
Observation 3: There's no clear CTA or timeline. Add one.
Rewritten: "Let's simplify your team's next product launch. I can walk you through how we cut 40% of the cycle in 2 weeks. Let me know if you'd like a 15-min walkthrough."

---

10. INDUSTRY FIT & REAL USE CASES

Ideal Industries:

- Solopreneurs & Consultants
- Founders & Fundraising Teams
- Strategic Writing / Public Thought Leaders
- High-Stakes Communication (Legal, Political, Corporate)
- Personal Brand Development

Real Use Cases:

- Editing personal newsletters or LinkedIn posts
- Simulating smarter investor pitch decisions
- Rewriting a resignation or hiring letter
- Practicing a founder AMA
- Debriefing a big mistake with clarity, not shame

---

11. ADVANCED LAYERS & EXTENSIONS

- **Emotion Dialing:** Create "You at 100% calm" or "You at 80% confident but cautious"

- **Clone vs. Critic Debates:** Simulate you and your toughest critic side-by-side
- **Future-Mirror Layer:** Ask: "What would I regret not seeing 6 months from now?"
- **Trust Mirror:** Create a clone that advises you *as if your legacy depends on it*
- **Voice-Only Rewriter:** Let AI rewrite your draft keeping your rhythm, but upgrading clarity
- **Ask AI to create *contrarian analogies*** to test if your goal can survive flipped logic.

E.g., "How would an anti-coach describe this framework?"

---

12. VALUE SUMMARY

User Level: Beginner
Usefulness: Provides guided thinking support. Helps you refine personal tone and decision clarity.

User Level: Intermediate
Usefulness: Allows self-critique and smarter editing with emotional distance.

User Level: Expert
Usefulness: Enables self-dialogue simulation for high-performance thinking.

User Level: CTO / Strategist / Consultant
Usefulness: Rapid internal coaching tool. Improves leadership messaging, investor logic, and feedback synthesis.

User Level: Educator / Coach
Usefulness: Trains reflection, clarity, and strategic self-talk via simulation.

---

13. LAW SNAPSHOT

Role: Inner Strategist Simulator
Lens: Metacognition, narrative control, self-alignment
Strength: Self-awareness, persuasive clarity, resilience
Weakness: Risk of self-confirmation if mirror is too flattering —
simulate honest logic, not ego

PRO TIPS & IDEAS TO TRY

- Ask AI to "simulate how my best coach would reframe this."
- Use this law before writing anything "from the gut" that will be seen publicly.
- Use post-failure: "What would wiser-me tell me to take from this experience?"
- Even better: "How would wiser-me explain this to a team I lead?"
- Use this law for annual reviews or career retrospectives: simulate your next-year-self advising you now."
- Use contrast prompts: "Explain this as a villain, then as a hero" — and compare tone and emphasis.

IDEAS TO TRY

- Run a Past-Future Reflection Chain:
1. Prompt: *"You're me, 1 year after I made this decision. What do you know now?"*
2. Then flip it: *"You're me from last year. What would you fear about this plan?"*
3. Compare answers. Insight lives in the tension.
- Add a Regret Filter:

"I'm you, post-regret. What advice do you wish I had followed earlier?"

• Use this law before big personal or business pivots: Simulate your future self giving a 5-bullet keynote titled: *"What I Learned From the Year I Almost Failed."*

🔗 PROMPT CHAIN LINK

Where this law connects to others. Use together for amplified results.

• Law 6: Cognitive Clone Critique — use both laws together: one as the smarter peer, one as your wiser future.
• Law 4: Prompt Memory Layering — mirror-self prompts get better if the AI remembers your past inputs or persona evolution.
• Law 11: Ethical Inversion Prompt — challenge your mirror-self to flip values and question whether your ethics are reactive or predictive.
• Law 17: Recursive Prompt Evolution — use Mirror Self to review and upgrade your prompt systems like a versioned software release.

---

14. LAW 9: MIRROR SELF PROTOCOL – [THE AI DECISION COMPASS] FINAL TAKEAWAY

The clearest advice often comes from the voice you haven't simulated yet.
Let AI become your strategist — by becoming the version of you that sees past emotion.

LAW 10: Echo Chamber Breaker [Bias Reversal Prompting]

To build fairer, sharper AI — challenge its echo, don't amplify it.

---

1. LAW SUMMARY

This law teaches how to break out of the AI's built-in confirmation patterns by deliberately reversing its bias trajectory. It trains you to prompt in ways that challenge assumptions, expand debate, and expose unseen angles — especially in high-stakes or sensitive topics.

---

2. METAPHORS / QUANTUM SLOGANS

- *If the room only echoes — leave the room.*
- *You don't correct bias by feeding it. You flip it.*
- *Ask AI to argue with itself — or it'll just agree with you.*
- *No reflection is perfect until you distort it on purpose.*
- *Good logic doesn't come from safe answers.*

---

3. MINI MANIFESTO / HOOK

You're not in an echo chamber — until you are.
Most prompt failures stem from trapped assumptions.
This Law is the escape hatch: Break the loop. Let new ideas in.

We unknowingly create echo chambers in our own questions — and the AI amplifies what it thinks we want.

This law stops that spiral.
By using "bias reversal prompting," we make the AI argue from different political, social, ethical, or commercial lenses.
Not to confuse it — but to stress-test it.
Because the only way to find balance... is to tip the scale on purpose.

---

4. FOUNDATIONAL IDEA (HUMAN SIDE + AI SIDE)

Human Side:
We all suffer from confirmation bias — seeking evidence that supports what we already believe. AI magnifies that when we don't check ourselves.
This law adds *cognitive friction* into the prompting process: reversing position, swapping ideology, simulating counterculture — so we can see blind spots, not just reflections.

AI Side:
LLMs generate based on statistical relevance — not moral correctness.
If your prompt leans a certain way, it will likely reinforce that angle.
Bias reversal prompting forces the model into a new reasoning track — generating more balanced, multidimensional responses.

---

5. CORE IDEA (EXECUTIVE SUMMARY)

Echo Chamber Breaker flips the default polarity of the model — especially around topics with high cultural, legal, or personal sensitivity.
By asking for the *opposite*, the *minority view*, or the *ethical*

counterweight, you generate contrast that sharpens clarity, reveals risks, and improves trust.

This is especially powerful in:

- Diversity-sensitive content
- Risk communications
- Policy writing
- Investor messaging
- Political analysis
- Global brand strategy

---

6. WHAT IT IS (FOUNDATIONAL CONCEPT)

Bias Reversal Prompting means asking the AI to:

- Simulate the opposite worldview
- Represent an underrepresented voice
- Challenge its own default framing
- Reframe from a different stakeholder's ethics
- Identify and explain *its own potential blind spots*

Types of reversals:

- **Ideological** (left vs. right)
- **Role-based** (CEO vs. intern)
- **Global** (Western vs. Eastern values)
- **Consumer vs. Producer**
- **Privacy vs. Innovation Lens**
- **Profit vs. Purpose Lens**

---

7. HOW TO USE IT (STEP-BY-STEP)

1. **Identify Your Original Prompt**
What's the default angle? Who's being centered? (e.g. "How should we launch this AI in healthcare?")
2. **Ask for a Reversal**
"What would a privacy ethicist challenge here?" or "How would a patient rights advocate view this?"
3. **Request Reasoned Opposition**
Don't just flip the emotion — ask for the opposing *logic*.
4. **Compare Both Outputs**
Analyze the divergence. Which framing is more thoughtful, balanced, or inclusive?
5. **Ask for a Synthesis**
Create a final version that integrates the best of both frames — or choose the more ethical direction.

---

8. PROMPT TEMPLATE (COPY-READY, MODULAR FORMAT)

SCENARIO: We're creating onboarding flows for a new AI productivity tool.
DEFAULT VIEW: The company wants to emphasize speed, automation, and scale.
BIAS REVERSAL PROMPT: Simulate the POV of a data privacy expert who distrusts rapid automation. What parts of this message would raise concerns, and how could we revise for transparency and ethical balance?

---

9. PROMPT OUTPUT EXAMPLE

88

"A privacy expert would flag the vague data policies and lack of opt-out controls. The phrase 'seamless integration' suggests background data scraping. Suggest rewording to: 'You control what gets automated — and what stays private.' Transparency about user consent is key."

---

10. INDUSTRY FIT & REAL USE CASES

Ideal Industries:

- Policy & Legal
- Tech & AI Product Development
- International Marketing
- Journalism & Media
- Corporate Governance
- Compliance

Real Use Cases:

- Writing gender-neutral hiring descriptions
- Reframing ad copy to avoid cultural stereotypes
- Testing product messaging across conservative vs. progressive audiences
- Avoiding bias in automated decision tools (e.g. HR, lending, housing)
- Running PR copy through ethics advisory simulation

---

11. ADVANCED LAYERS & EXTENSIONS

Self-Diagnosis Prompt: Ask AI: "Where might your response reflect unconscious bias?"

Multi-Culture Frames: Request 2–3 global takes: American, Indian, Brazilian, Japanese.

Values Framing: Rewrite from the lens of "data dignity" vs. "data profit".

Double-Reversal Loop: Run both left- and right-leaning frames, then synthesize.

Red Team Prompting: Simulate activist critiques or public controversy simulations

---

12. VALUE SUMMARY

User Level: Beginner
Usefulness: Introduces the idea that AI is not neutral — and helps find blind spots.

User Level: Intermediate
Usefulness: Improves fairness, accuracy, and trustworthiness in messaging.

User Level: Expert
Usefulness: Allows simulation of opposing stakeholders for policy-proofing and audience sensitivity.

User Level: CTO / Strategist / Consultant
Usefulness: Prevents reputational risk and builds credibility in AI-powered communications.

User Level: Educator
Usefulness: Teaches responsible AI use, bias identification, and ethics modeling.

---

13. LAW SNAPSHOT

Role: Ethical Challenger
Lens: Polarity, role-flipping, moral contrast
Strength: Fairness, depth, transparency
Weakness: Can cause confusion if prompts are vague — requires clarity on "which voice" to simulate

---

PRO TIPS & IDEAS TO TRY

- *Ask: "What would an ethics board reject in this message?"*
- *Use this law with legal teams before any AI-generated public copy goes live*
- *Teach teams to create two versions of messaging: default and reversed, then A/B test impact*
- *Combine with Law 5 (Lens Reversal) to sharpen both creativity and safety*
- *Ask: 'Which assumptions in this message are too agreeable — and to whom?'*

IDEAS TO TRY

Use the **Adversarial Reviewer Frame**:

"Pretend you're the person who disagrees with every word I just wrote. Where are the weak points?"
Run this before publishing or sending.

Try a **Public vs. Private Test**:

"Give me the honest opinion a stakeholder might say behind closed doors — not in front of the team."

Simulate **2+ Bias Reversals**:

"Argue this from the POV of a skeptical CFO, then from a passionate activist. Where do their concerns overlap?"

Use **moral contrasts**:

"What if this policy is ethical for me but harmful for someone in a weaker position? How should I frame it then?"

🔗 PROMPT CHAIN LINK

Where this law connects to others. Use together for amplified results.

Law 13: Robustness Testing — challenge the logic and moral frame under fire before it reaches critics.
Law 6: Cognitive Clone Critique — compare clone-self output with an ideological adversary's POV to find blind spots.
Law 11: Ethical Inversion Prompt — go deeper by reversing not just belief, but the moral assumption behind it.
Law 9: Mirror Self Protocol — ask: "Would future-me regret not challenging this viewpoint harder?"
Law 12: Silence Amplification — once bias is revealed, ask what topics or truths were excluded from the discussion.

---

14. LAW 10: ECHO CHAMBER BREAKER [BIAS REVERSAL PROMPTING] FINAL TAKEAWAY

The most responsible AI users don't just prompt for clarity. **They prompt for challenge — because progress happens when assumptions get flipped.**

LAW 10 BONUS – SIMULATED COLLABORATION

"When you simulate a smart partner, you stop working alone."

Insightful Tip:
Use role-based prompting to simulate experts, editors, critics — even competitors. It unlocks angles you'd never think of alone.

Real-World Prompt Flip:

"Play two roles: startup founder and skeptical investor. Have a 5-message conversation about this pitch."

Mini-Mindset Shift:
The best prompting isn't solo.
It's improv.

LAW 11: Ethical Inversion Prompt [See the Opposite Truth]

Sometimes the smartest answer is the one that makes you uncomfortable.

---

1. LAW SUMMARY

This law teaches how to use moral inversion — asking AI to argue against the "good" outcome or simulate the downside of a beloved idea. By flipping ethics on their head, you reveal hidden risks, emotional blind spots, and ethical tradeoffs your default lens may ignore.

---

2. METAPHORS / QUANTUM SLOGANS

- *Ethics aren't one-directional. Neither should your prompts be.*
- *Before you act with conviction — test with inversion.*
- *The devil's advocate wears a halo in the right light.*
- *To see the full shape of an idea, cast a shadow behind it.*
- *What seems good from here… may look selfish from over there.*

---

3. MINI MANIFESTO / HOOK

Most people use AI to confirm their ethics.
But the truly strategic — and responsible — ask it to *test* those ethics.

This law lets you simulate the exact opposite of your default "good":

- Is speed always good in hiring?
- Is scale always good in tech?
- Is personalization always good in health?

Not because you want to sabotage your own idea — but because *moral clarity comes from friction.*
Great thinkers build ideas by debating their shadows.

---

4. FOUNDATIONAL IDEA (HUMAN SIDE + AI SIDE)

Duality isn't just moral — it's tactical. By simulating both applause and outrage, you gain early warning radar against PR crises or political backlash.

Human Side:
We often fall for *noble-sounding fallacies* — good ideas that become harmful at scale, in the wrong context, or when misused.
This law pulls from philosophy, critical ethics, and "steelman" argumentation — where the best argument for the opposing side is explored before choosing your own.

AI Side:
LLMs often default to "positive outputs" when prompted with feel-good framing.
But when deliberately inverted — "What's the hidden cost of this helpful idea?" — the model can generate far more mature, nuanced insights.
This is where it becomes an ethical thinking partner, not just a script generator.

---

5. CORE IDEA (EXECUTIVE SUMMARY)

Ethical Inversion Prompting is the practice of testing your best intentions by asking AI to simulate their unintended consequences — morally, socially, psychologically, or legally.

It's not negativity. It's **strategic discomfort**.
And it's the fastest way to:

- Prevent future scandals
- Avoid culture clashes
- Detect power asymmetries
- Refine your brand trust

---

6. WHAT IT IS (FOUNDATIONAL CONCEPT)

This law creates **moral complexity** by flipping assumptions.
You ask AI to explore:

What harm might this "good thing" cause at scale?
What belief system would reject this approach?
What's the unintended consequence of pursuing this ideal?
What happens if the wrong people misuse this helpful tool?

It borrows from:

Design Ethics
Dark Pattern Recognition
Behavioral Economics
Adversarial Role Simulation

---

7. HOW TO USE IT (STEP-BY-STEP)

1. **Start with a "Good" Intention or Strategy**
What do you believe is ethical, inclusive, or helpful?
2. **Frame a Reversal Challenge**
Ask: "Who might see this as harmful — and why?"
3. **Prompt AI to Simulate That Opposing Belief System**
Have it argue with reason, not rage.
4. **Request a Risk Map**
Ask AI to show 3 possible harms that could arise unintentionally.
5. **Synthesize With Care**
Extract the insight. Refine your strategy, tone, or action plan accordingly.

---

8. PROMPT TEMPLATE (COPY-READY, MODULAR FORMAT)

SCENARIO: Our startup offers AI tools to improve student performance by real-time learning feedback.
ASSUMED GOOD: This empowers students to learn faster and close gaps.
ETHICAL INVERSION: Simulate the perspective of a parent or psychologist concerned with student privacy, self-worth, and AI pressure.
TASK: What 3 ethical risks could arise? How might we reframe the product to reduce those risks while preserving its benefits?

---

9. PROMPT OUTPUT EXAMPLE

"1) Constant feedback may cause performance anxiety or perfectionism in younger students.
2) AI-driven personalization could override teacher autonomy.
3) Privacy concerns if sensitive data is used for adaptive learning without full consent. To reduce risk: emphasize teacher-AI collaboration, include opt-in control, and reframe benefits as student-guided, not performance-optimized."

10. INDUSTRY FIT & REAL USE CASES

Ideal Industries:

- EdTech
- AI-Powered Healthcare
- HR & Recruiting
- FinTech
- Government / Civic Tech
- Legal Technology

Real Use Cases:

- Preventing algorithmic discrimination in hiring tools
- Reframing mental health apps to avoid "toxic productivity"
- Reviewing ethical downsides of hyper-personalized ads
- Challenging surveillance culture in smart cities
- Testing cultural backlash scenarios before policy rollouts

11. ADVANCED LAYERS & EXTENSIONS

Temporal Inversion: Ask: "What might future historians criticize about this idea?"

Role-Flipped Feedback: Run the same idea through activist, journalist, whistleblower personas

Ethics Layer Generator: Ask AI to generate 3 conflicting ethical values and force a decision scenario

Contextual Reversal Loop: Simulate the same message across regions or social groups to detect mismatches

Legacy Prompting: Ask: "What might we regret about this message in 10 years?"

Ethical Checkpoint
Before publishing or sharing any AI-generated content, ask:
"Who might object to this — and why?"
This reduces tone-deaf launches and increases public trust.

---

12. VALUE SUMMARY

User Level: Beginner
Usefulness: Introduces complexity and ethical reflection early in AI use.
User Level: Intermediate
Usefulness: Helps prevent PR and policy mistakes by revealing unintended consequences.
User Level: Expert
Usefulness: Enables nuanced reframing, long-term impact modeling, and ethical tradeoff analysis.
User Level: CTO / Strategist / Consultant
Usefulness: Provides future-proofing tools for responsible innovation, trust building, and brand integrity.
User Level: Educator

Usefulness: Trains students to challenge their own "best" ideas with empathy and critical reasoning.

---

13. LAW SNAPSHOT

Role: Ethical Contrarian / Inversion Analyst
Lens: Strategic risk, unintended consequence, moral reframing
Strength: Maturity, foresight, cultural intelligence
Weakness: Can cause discomfort or indecision without structured synthesis

---

PRO TIPS

Before launching any "do-good" campaign, run an inversion prompt first

Pair this with Law 10 (Echo Chamber Breaker) to test both social and ethical polarities

Use for compliance or DEI review processes to reduce tone-deaf messaging

Ask AI: "If this backfires ethically — what headline might appear?"

Simulate a protestor's response to your message. Not to agree — but to prepare.

IDEAS TO TRY

Run the **Hero/Villain Test**:
"In this story, I'm the hero. Rewrite this from the villain's perspective who thinks they're doing the right thing."
Great for sales, politics, brand messaging, or ethics training.

Use the **Moral Time Machine**:
"If this decision were made 50 years ago, would it have been praised or condemned? What changed?"
Helps detect ethics that are situational or trending, not grounded.

Try the **Stakeholder Reversal Prompt**:
"If this benefits my department but harms a group we never hear from, what's the ethical blind spot?"

PROMPT CHAIN LINK

Where this law connects to others. Use together for amplified results.

Law 10: Echo Chamber Breaker — first reveal surface-level bias, then invert the ethical foundation behind it.
Law 14: Ethical Duality Simulation — explore multiple "truths" side-by-side with full emotional and moral context.
Law 13: Robustness Testing — put your values under pressure, then flip them to simulate worst-case failures.
Law 7: Inverted Industry Metaphor — use upside-down framing to reimagine moral dilemmas through unfamiliar lenses (e.g., What would a nonprofit CEO do?).

14. LAW 11: ETHICAL INVERSION PROMPT [SEE THE OPPOSITE TRUTH] FINAL TAKEAWAY

The most dangerous assumptions are the ones that feel too noble to question.

Invert the light — and you may see the shadow that saves your integrity.

LAW 12: Silence Amplification [Prompting What's Missing]

What's absent from the answer often matters more than what's said.

---

1. LAW SUMMARY

This law teaches you how to detect and fill blind spots in AI output. Rather than improving what's visible, it prompts the model to reveal what was left unsaid — the missing context, questions, voices, and alternatives that deepen truth and improve trust.

---

2. METAPHORS / QUANTUM SLOGANS

- *Silence is an answer — if you know how to listen.*
- *The missing piece isn't noise. It's the shape of the truth.*
- *Great prompting isn't what you ask. It's what you notice wasn't answered.*
- *Fill the white space, not the margins.*
- *What the AI didn't say... is often your biggest risk.*

---

3. MINI MANIFESTO / HOOK

When AI responds quickly, we're often impressed. But we rarely ask: *What did it skip?*

What point of view, stakeholder, risk, constraint, or deeper issue never got mentioned?

This law isn't about better writing. It's about **silent insight recovery** — finding the wisdom between the words.
You prompt the model to expose what it missed. Not by guessing, but by listening for absence.
Silence is a prompt. Missing is a message. Ask for it.

---

4. FOUNDATIONAL IDEA (HUMAN SIDE + AI SIDE)

Human Side:
We often trust what's said — and ignore what isn't. Yet silence shapes decisions, just like speech. Think of unmentioned risks in marketing, unstated assumptions in politics, or who's not "in the room" when plans are made.

AI Side:
LLMs generate based on probability — which means the most common, expected, or obvious answers rise first.
The uncommon, uncomfortable, or subtle ideas get cut.
But they're still in the model — you just have to **ask what it skipped.**

---

5. CORE IDEA (EXECUTIVE SUMMARY)

Silence Amplification is the deliberate act of asking AI:

- *What did you leave out?*
- *Who wasn't represented in that answer?*
- *What angle was ignored — and why?*
- *What deeper truths are hiding behind this simplification?*

104

When used strategically, this law reveals:

- Cultural bias
- Stakeholder gaps
- Risk underrepresentation
- False consensus
- Missed emotional resonance

---

6. WHAT IT IS (FOUNDATIONAL CONCEPT)

This technique is about simulating *omission awareness*. You're not just asking "what else?" — you're asking:

- *"What group might feel left out by this?"*
- *"What risk did I fail to name?"*
- *"Which stakeholder's concerns are unstated?"*
- *"What deeper layer of meaning is implied but unsaid?"*

Forms of silence prompting:

- **Cultural Voicing** – "What cultures might misread this message?"
- **Assumption Excavation** – "What assumptions were made without being named?"
- **Omission Audit** – "List 3 relevant things that were not mentioned — but matter."
- **Value Gap Prompt** – "What important ethical value might be missing from this plan?"

---

6. HOW TO USE IT (STEP-BY-STEP)

What to avoid: Don't let the AI "mirror" you too much. If it becomes your cheerleader, it loses critique power. Push it to challenge, not echo.

1.   **Run Your Original Prompt**
Get your normal output — a plan, message, idea, list.
2.   **Ask for an Omission Audit**
"Now tell me 3 things that were left out — but might be important to someone else."
3.   **Request Silent Stakeholder Simulation**
"Who would feel unseen or unrepresented in this?"
4.   **Generate a Revised Output That Includes the Missing**
Ask AI to rewrite the original — this time *including the omissions* it just uncovered.
5.   **Evaluate Depth and Inclusion**
Did the message gain richness, fairness, or clarity? Use this to improve final version.

---

8. PROMPT TEMPLATE (COPY-READY, MODULAR FORMAT)

SCENARIO: I wrote an onboarding email for our AI hiring platform. It emphasizes speed, automation, and ease.
SILENCE AMPLIFICATION:

- What perspectives are missing?
- Who might feel left out or skeptical?
- What key value or risk was underrepresented?

Then rewrite a version that includes those voices — without bloating the message.

---

9. PROMPT OUTPUT EXAMPLE

"Missing perspective: The recruiter's fear of losing human judgment. Unmentioned risk: Bias amplification. Emotional gap: Candidate anxiety. Rewritten: 'Our AI supports — not replaces — your human instincts. Stay in control while hiring faster, fairer, and with more insight.'"

---

10. INDUSTRY FIT & REAL USE CASES

Ideal Industries:

- HR Tech
- Healthcare Communications
- Public Policy
- Multicultural Marketing
- DEI Strategy
- Legal, Risk & Compliance

Real Use Cases:

- Rewriting HR docs to be more inclusive
- Testing PR messages for who they exclude
- Finding "ghost values" in policy proposals
- Reframing tech onboarding for non-experts
- Auditing AI-generated training content for blind spots

---

11. ADVANCED LAYERS & EXTENSIONS

- **Time-Based Omission:** Ask: "What will this message age poorly on in 2 years?"

- **Trust Layer Check:** "What reassurance is missing that would build trust?"
- **Silence Comparison Prompt:** Compare version A (fast and polished) vs. version B (slower but inclusive)
- **Voice Swap:** Simulate response as if written by someone who feels excluded
- **Emotional Omissions:** "What unspoken emotions might this message ignore?"

---

12. VALUE SUMMARY

User Level: Beginner
Usefulness: Introduces awareness of what AI skips — and why it matters.

User Level: Intermediate
Usefulness: Teaches how to prompt for emotional depth and missing details.

User Level: Expert
Usefulness: Enables layered quality control — ethical, emotional, strategic.

User Level: CTO / Strategist / Consultant
Usefulness: Critical for stakeholder alignment, legal defensibility, and brand trustworthiness.

User Level: Educator
Usefulness: Helps students learn critical reading, revision, and inclusion principles via AI reflection.

---

13. LAW SNAPSHOT

Role: Inclusion Auditor / Silent Risk Mapper
Lens: Negative space, unsaid truth, stakeholder simulation
Strength: Depth, fairness, reflection
Weakness: Can slow output if overused; must balance insight with clarity

---

PRO TIPS

- Ask AI to "rewrite this message as if a forgotten stakeholder finally spoke up."
- Run a Silence Amplification prompt on your own resume, pitch, or proposal
- Use as final QA step before publishing or shipping any high-stakes message
- Pair with Law 11 for ethical + silence-based double-check
- Simulate a 2-person AI loop: One friendly, one hostile.
Let them take turns evaluating the same draft — then combine feedback.

IDEAS TO TRY

- Use the **"Missing Risk" Prompt**:

*"This seems solid — but what is **conspicuously absent** that could cause reputational, legal, or ethical risk?"*

- Try a **Pre-Mortem Gap Scan**:

*"Assume this idea failed in the real world. What was **left out** that caused the failure?"*

- Run a **2-column prompt test**:
1.   *"List what this message includes."*
2.   *"Now list what someone might expect — but won't find."*
Compare for alignment.
- Add an **emotional silence layer**:

"What feelings, fears, or questions might this ignore — especially among skeptics or underrepresented groups?"

🔗 PROMPT CHAIN LINK

Where this law connects to others. Use together for amplified results.

- **Law 11: Ethical Inversion** — after flipping ethical assumptions, use this law to detect the values or harms left unspoken.
- **Law 10: Echo Chamber Breaker** — break groupthink first, then identify what the echo chamber never even discusses.
- **Law 3: Prompt Autopsy & Debugging** — if something keeps failing subtly, use this law to detect structural absences.
- **Law 16: Multi-AI Role Rotation** — assign different AIs to "speak for the silenced" — the persona, stakeholder, or customer often forgotten.

14.
LAW 12: SILENCE AMPLIFICATION [PROMPTING WHAT'S MISSING] FINAL TAKEAWAY

What the AI forgets to say is often what your audience most needs to hear.

Don't just fix what's flawed. Find what was never spoken — and amplify it.

LAW 13: Robustness Testing via Adversarial Framing [Break It Before They Do]

If your AI idea can't survive its harshest critic — it's not ready.

1. LAW SUMMARY

This law transforms prompting into a stress test. Instead of building ideas in a vacuum, you simulate an adversary — a competitor, regulator, or activist — and ask them to attack your idea, messaging, or strategy. If it breaks under pressure, it wasn't ready for reality.

2. METAPHORS / QUANTUM SLOGANS

- *Don't ship until you shatter it.*
- *Break your idea — so the market can't.*
- *If it can't survive a hostile question, it's not a strategic answer.*
- *AI gives you the mirror. Now show it the threat.*
- *A perfect pitch is the one that wins even after critique.*

3. MINI MANIFESTO / HOOK

Most people use AI for creation — not confrontation.
They ask it to build, polish, promote. But rarely: *destroy.*

This law changes that.

Robustness testing is adversarial prompting done on purpose. You simulate backlash, resistance, or competitive teardown to see where your logic, message, or model will crack.
You ask: *If someone wanted to tear this down — how would they?*
Then: *How do I fix that first?*

Great founders, strategists, and communicators don't fear critique. They **simulate it first.**

---

4. FOUNDATIONAL IDEA (HUMAN SIDE + AI SIDE)

Human Side:
We're wired to avoid criticism — especially before launching something we believe in. But in military, policy, and negotiation — red-teaming is a standard practice. It strengthens resilience through **preemptive pressure.**

AI Side:
LLMs can simulate highly specific adversaries — an angry customer, a skeptical investor, a hostile journalist — and surface attack vectors you hadn't considered.
This doesn't weaken your idea. It future-proofs it.

---

5. CORE IDEA (EXECUTIVE SUMMARY)

This law turns your AI into a stress-test generator.
It's not just: *"How can I make this better?"*
It's: *"How would someone try to destroy this?"*
Then: *"How do I make it unbreakable?"*

Robustness prompting is critical for:

113

- High-stakes communication
- Investor and boardroom prep
- Product launches
- Legal compliance
- Crisis planning
- Leadership messaging

---

6. WHAT IT IS (FOUNDATIONAL CONCEPT)

This law builds on **red team logic**, adapted for solo AI collaboration. You prompt AI to become:

- The *toughest critic*
- A *ruthless competitor*
- A *skeptical investor*
- A *disillusioned user*
- A *regulatory watchdog*

Then ask it to:

- Expose weaknesses
- Question assumptions
- Attack from an emotional, ethical, or financial angle
- Offer stronger alternatives

---

7. HOW TO USE IT (STEP-BY-STEP)

1. **Define What You Want to Stress-Test**
Is it a message, strategy, product, vision, or hiring plan?
2. **Choose the Adversarial Persona**
Who's the most dangerous critic? Simulate that role.

114

3.   **Prompt AI to Attack**
Use language like: "What would [critic] say to dismantle this idea?"
4.   **Identify 3–5 Weak Points**
Ask AI to rank the risks — from most to least dangerous.
5.   **Reinforce or Redesign**
Prompt AI to rephrase, rebuild, or defend the idea from each attack.

---

8. PROMPT TEMPLATE (COPY-READY, MODULAR FORMAT)

SCENARIO: I'm pitching a new AI co-pilot for small business accounting.
TASK: Simulate a skeptical CFO at a mid-sized firm. What objections, fears, or criticisms might they have about this tool? List 5 weaknesses or concerns. Then, rewrite the core pitch to preemptively address them.

---

9. PROMPT OUTPUT EXAMPLE

Objections:

1.   Data security: Who has access to sensitive financials?
2.   Accuracy: What if the AI misclassifies expenses?
3.   Regulatory risk: Will this comply with changing tax laws?
4.   Accountability: Who's liable for errors?
5.   Vendor trust: Too many tools already — why trust yours?

Rewritten pitch:
"Built for CFOs who care about precision and control. Our AI co-pilot runs securely on encrypted servers, flags anomalies in real time, and

115

stays ahead of tax updates — while keeping your team in command of every final approval."

---

10. INDUSTRY FIT & REAL USE CASES

Ideal Industries:

- Cybersecurity
- Health Tech & Med Devices
- FinTech
- LegalTech
- SaaS & Product Strategy
- Politics & Public Affairs

Real Use Cases:

- Pitch testing before investor meetings
- Messaging review before crisis response
- Product positioning review for skeptical customers
- Testing terms of service before legal review
- Simulating hostile media questions before launch

---

11. ADVANCED LAYERS & EXTENSIONS

Multi-Attack Simulation: Run 3 adversaries at once — user, lawyer, and competitor

Emotional Attack Layer: Ask AI to simulate outrage, betrayal, or fear

Cultural Critic Prompt: "How would someone from a different cultural norm view this?"

Legacy Check: "What might a scandal headline say if this idea backfires?"

Adversary-to-Advisor Loop: Once attacked, flip role and ask: "How would this critic help fix it?"

Run *Anticipation Prompts*:

"If this output fails in public — what's the headline?"
Then rewrite with that failure in mind.

---

12. VALUE SUMMARY

User Level: Beginner
Usefulness: Prepares you to spot the flaws before launch.
User Level: Intermediate
Usefulness: Adds strategic realism to messaging and decision-making.
User Level: Expert
Usefulness: Builds resilience into high-stakes workflows and public-facing projects.
User Level: CTO / Strategist / Consultant
Usefulness: Critical for board-level defense, investor readiness, and public trust strategies.
User Level: Educator
Usefulness: Teaches strategic debate, self-critique, and adversarial resilience skills.

---

13. LAW SNAPSHOT

Role: Red Team Simulator
Lens: Adversarial framing, scenario reversal, attack-resilience
Strength: Strategic robustness, risk anticipation
Weakness: Can create paranoia or over-engineering if not used with final synthesis

CROSS-LINK

Use with Law 11: Ethical Duality Simulation to prepare for worst-case reactions.

PRO TIPS & IDEAS TO TRY

Use this law before any major launch, not after pushback starts.
Have AI simulate a journalist covering a scandal based on your current copy
Rebuild messaging after critique — and A/B test before going live
Train junior teams to use this law as part of QA, legal, or comms review
Run this law weekly with your sales copy, onboarding flows, or pitch decks — and you'll always be ahead of your critics.

IDEAS TO TRY

Use the Hostile Stakeholder Prompt:
"You're the compliance officer who doesn't trust this language. What would you flag — and why?"

Try the "Sabotage Frame":
"You are trying to make this prompt fail. Where would you exploit it, confuse it, or cause hallucination?"

Run a Pre-Publication Attack Simulation:
"Pretend this article is being torn apart by someone on Twitter who hates the idea. What would they mock, question, or attack?"

Combine Robustness + Silence Amplification:
After simulating failure, ask: *"What did I fail to anticipate?"*

You'll surface vulnerabilities most authors, marketers, and executives never test.

🔗 PROMPT CHAIN LINK

Where this law connects to others. Use together for amplified results.

Law 12: Silence Amplification — after simulating failure, detect what was absent that enabled it.
Law 10: Echo Chamber Breaker — stress-test your logic by simulating an ideological or political opponent.
Law 3: Prompt Autopsy & Debugging — use after robustness testing to clean up failed structures and simplify confusion.
Law 17: Recursive Prompt Evolution — use failure data to create the next, stronger version of your prompt stack.
Law 4: Prompt Memory Layering — ensure robustness across steps by embedding failure-prevention logic in memory loops.

14. LAW 13: ROBUSTNESS TESTING VIA ADVERSARIAL FRAMING [BREAK IT BEFORE THEY DO] FINAL TAKEAWAY

The strongest strategy is the one that already survived its critics — in simulation.
If you can break it in private, no one can break it in public.

LAW 14: Ethical Duality Simulation [Truth vs. Loyalty Framing]

Sometimes what's right and what's loyal aren't the same — prompt for both.

---

1. LAW SUMMARY

This law explores how to prompt AI to navigate the tension between truth and loyalty — between objective facts and personal, brand, team, or cultural alignment. You simulate both ethical poles to see how AI can honor facts without betraying trust, or challenge bias without breaking rapport.

---

2. METAPHORS / QUANTUM SLOGANS

- *Facts are data. Loyalty is survival.*
- *The truth doesn't always fit inside a brand guide.*
- *Don't force AI to pick sides. Ask it to simulate both.*
- *Between truth and loyalty lives diplomacy.*
- *Your smartest prompt isn't who's right — it's who gets heard.*

---

3. MINI MANIFESTO / HOOK

Most people ask: "What do I want *now*?"
But the greatest thinkers ask: *"What would my future self want me to*

120

ask?"
This Law unlocks temporal vision in every prompt.

Your customer is wrong — but loyal.
Your team loves a plan that might fail.
Your brand claims fairness — but your algorithm shows bias.

What do you say? Who do you protect? What do you prompt for?

This law invites you to simulate the dilemma — not dodge it.
You ask AI to respond from both lenses: *loyalist* and *truth-teller*.
And then? You synthesize — or choose your line.

Because great leadership doesn't mean always being honest or always
being loyal.
It means **knowing when and how to honor both.**

---

4. FOUNDATIONAL IDEA (HUMAN SIDE + AI SIDE)

Human Side:
In real life, we balance facts with feelings, data with loyalty, and
honesty with hierarchy. Truth can be alienating. Loyalty can be
blinding.
This law builds on moral philosophy, diplomacy, and stakeholder
psychology.

AI Side:
LLMs default to tone-matching — which can either blindly reinforce
groupthink or offer robotic truth. But by prompting both **truth
framing** and **loyalty framing**, you reveal underlying tensions — and
find smarter, safer middle ground.

---

5. CORE IDEA (EXECUTIVE SUMMARY)

Ethical Duality Simulation prompts AI to articulate **both sides of moral tension**:

- The version that tells the truth bluntly
- The version that preserves group cohesion, emotional safety, or institutional trust

It's about **simulating the ethical dilemma** — not solving it prematurely.
This law is critical for:

- Internal messaging
- Crisis communication
- Founder leadership
- Whistleblowing scenarios
- Cultural intelligence

---

6. WHAT IT IS (FOUNDATIONAL CONCEPT)

This technique simulates **dual perspectives**:

- *Truth Lens:* Prioritizes objectivity, accuracy, long-term impact
- *Loyalty Lens:* Prioritizes social bond, team harmony, reputation, or legacy

It forces moral clarity by surfacing tension — not erasing it.
Think of it as a decision lens for ethical nuance, not a binary switch.

Scenarios include:

- A marketing team pushing false urgency

- A founder asking for support after failure
- A hiring process masking flaws for optics
- A product team resisting critical feedback

---

7. HOW TO USE IT (STEP-BY-STEP)

1. **Define the Dilemma or Message**
What needs to be communicated, decided, or addressed?
2. **Prompt the Truth-Teller Version**
Ask AI to write the honest, factual, potentially uncomfortable version.
3. **Prompt the Loyalty-Preserving Version**
Ask AI to write the supportive, brand-aligned, or team-safe version.
4. **Compare Tone, Logic, and Risk**
Where do they align? Where do they clash?
5. **Synthesize or Choose Your Ethical Line**
You can blend both, or pick one — but now you're choosing with clarity.

---

8. PROMPT TEMPLATE (COPY-READY, MODULAR FORMAT)

SCENARIO: We're behind on our Q4 goals. The internal update must be sent to a 30-person team that trusts leadership, but hasn't heard about the setback.
TASK:

- Version 1: Write the honest, direct truth version — no sugarcoating.
- Version 2: Write the brand-loyal version that maintains morale and trust.

123

- Final: Synthesize a middle-ground version that communicates transparency *and* optimism.

---

9. PROMPT OUTPUT EXAMPLE

Truth Version:
"We're 17% behind target. A missed hiring cycle and client churn were key factors. If trends hold, bonuses may be impacted."
Loyalty Version:
"Despite a dynamic Q4, we've retained strong client engagement. Our team's resilience sets us up for a strong Q1."
Synthesis:
"We're currently behind target, but the cause is clear — and fixable. The next 30 days will define our Q1 momentum. Thank you for your trust — and your grit."

---

10. INDUSTRY FIT & REAL USE CASES

Ideal Industries:

- Corporate Leadership
- HR & Internal Comms
- Brand Management
- Healthcare Ethics
- Politics & Civic Tech
- Education & Nonprofit Strategy

Real Use Cases:

- Crisis communication (internal and external)
- Drafting letters from leadership

- Preparing public statements post-mistake
- Handling performance reviews with empathy
- Communicating ethical dilemmas in education or health

---

11. ADVANCED LAYERS & EXTENSIONS

- **Role Swaps:** Run both versions from the CEO, then from an intern
- **Impact Mapping:** Ask AI: "What's the short-term vs. long-term risk of each approach?"
- **Tone Bridging:** Simulate a 3rd party mediator's version of the message
- **Stakeholder Trust Test:** Ask: "Which version builds long-term loyalty, and why?"
- **Multi-Culture Lens:** How would this play across different cultures or power dynamics?

---

12. VALUE SUMMARY

User Level: Beginner
Usefulness: Helps navigate communication dilemmas with integrity.
User Level: Intermediate
Usefulness: Builds emotional intelligence into prompts, not just logic.
User Level: Expert
Usefulness: Enables leadership tone calibration and ethical storytelling.
User Level: CTO / Strategist / Consultant
Usefulness: Critical for crisis prep, reputation protection, and stakeholder clarity.
User Level: Educator

Usefulness: Teaches students how to hold moral tension in real-world scenarios.

13. LAW SNAPSHOT

Role: Ethical Negotiator / Tone Synthesizer
Lens: Truth vs. Loyalty, moral contrast, stakeholder empathy
Strength: Emotional precision, nuance, long-term trust
Weakness: Can produce diluted messaging if not purposefully synthesized

---

PRO TIPS & IDEAS TO TRY

Use this law for "pause moments" — when your gut says, "This message matters."

Always label the AI's responses clearly: Truth Version vs. Loyalty Version.

Revisit this law before sending anything on behalf of leadership, team, or mission.

Teach juniors to see tone as a strategic tool, not just a writing style. Let AI simulate both the hard truth and the emotionally safe version. Then have it write the synthesis.

Most ethical dilemmas aren't about right vs. wrong — they're about truth vs. loyalty. Simulate both, then explore the tradeoff.

Use this law when the right answer offends the team, the brand, or the institution. Let AI surface both paths.

Your job isn't to choose one frame — it's to understand the cost of

126

each.

If I had to live with this prompt for 10 years, would I still use it?
"How will this look in the past's eyes — or a future audit?"

IDEAS TO TRY

Run a Loyalist vs. Whistleblower Debate:

"Simulate a team member defending the company's image, and another defending the public's right to know. Who wins — and why?"

Try the Split Frame News Brief:

"Write a press release as if you were telling the uncomfortable truth."
Now write it as if your job depends on protecting the brand's reputation.
Then compare.

Add a Time Test:

"This may look loyal today — but how will it age in 5 years if the truth leaks? Rewrite for both versions."

Combine with Mirror Self Protocol:

"Will future-me respect the choice I made here, or wish I had told the harder truth sooner?"

PROMPT CHAIN LINK

Where this law connects to others. Use together for amplified results.

- **Law 11:** Ethical Inversion Prompt — simulate what happens when your truth becomes someone else's harm.
- **Law 6:** Cognitive Clone Critique — use your wiser self to judge both the loyal and the honest frame.
- **Law 13:** Robustness Testing — test how each ethical choice holds up under public pressure, social media backlash, or legal scrutiny.
- **Law 12:** Silence Amplification — ask: *"Which version hides the most?"* or *"What's ethically missing from each frame?"*

14. LAW 14: ETHICAL DUALITY SIMULATION [TRUTH VS. LOYALTY FRAMING] FINAL TAKEAWAY

AI can't decide your ethics — but it can show you both paths.
Simulate both voices. Speak with purpose. Choose with clarity.

LAW 14 BONUS – TIME & ENERGY BIASES

"AI doesn't forget. But it does get lazy."

Insightful Tip:
Prompting late in a long session can introduce drift. Use fresh sessions for important tasks — or re-anchor with context.

Real-World Prompt Flip:

"Restart thinking: Here's fresh context. Rebuild the plan from scratch in 3 steps."

Mini-Mindset Shift:
AI doesn't get tired.
But your signal gets fuzzy.

LAW 15: Format-for-Transfer Prompting [Train It to Travel]

Smart prompts aren't written for today's AI — they're built to survive tomorrow's.

---

1. LAW SUMMARY

This law teaches how to format prompts and AI outputs for easy transfer between models, teams, workflows, or future tools. By designing with clarity, structure, and interoperability in mind, you future-proof your prompting and make it scale-ready.

---

2. METAPHORS / QUANTUM SLOGANS

- *Prompt like it's code — not a conversation.*
- *What you format today becomes someone else's success tomorrow.*
- *Transferable prompts are the duct tape of intelligent systems.*
- *If your prompt only works here and now, it wasn't smart — it was lucky.*
- *Train your AI like it's going to travel without you.*

---

3. MINI MANIFESTO / HOOK

Most prompts are designed for a single use — and die there.
That's fine for casual tasks. But if you're building real workflows,
SOPs, or libraries, that approach **breaks under pressure**.

This law upgrades your prompt thinking from *situational* to *systemic*.
You stop writing for ChatGPT — and start writing for any AI, team
member, or future platform.

It's not just about clarity. It's about **survivability** — the ability of your
prompt to *travel*, be reused, updated, and understood without you
being in the room.

---

4. FOUNDATIONAL IDEA (HUMAN SIDE + AI SIDE)

Human Side:
Great systems designers build *for others*, not just for themselves.
Transferable prompts are like smart templates: clear inputs, flexible
outputs, and baked-in context — so others can run with them.

AI Side:
LLMs interpret prompt structure *better* than vague prose. When you
format prompts into clean roles, tasks, constraints, and output rules —
they perform better *and* can be transferred between models like GPT,
Claude, Gemini, or future AI agents.

---

5. CORE IDEA (EXECUTIVE SUMMARY)

Format-for-Transfer Prompting is the discipline of designing prompts
to **function across platforms, people, or future workflows**.
You break the prompt into modular, label-based sections (role,

context, task, constraints, examples, format). You minimize platform dependencies and maximize repeatability.

Result:
Your prompt doesn't just work — it **travels, scales, and survives**.

---

6. WHAT IT IS (FOUNDATIONAL CONCEPT)

This technique uses structured, modular prompt formatting — often resembling:

- Labeled sections: **ROLE, CONTEXT, TASK, FORMAT, EXAMPLES**
- Plain syntax: No model-specific slang
- Style constraints: Defined tone, output type, do/don't behavior
- Variable substitution: Inputs that can be swapped by teammates or scripts

You make it:

- Easy to update
- Easy to hand off
- Easy to track versions
- Easy to adapt for GPT, Claude, Gemini, or future agents

---

7. HOW TO USE IT (STEP-BY-STEP)

1. **Break Your Prompt Into Modular Parts**
Use headers like ROLE, TASK, CONTEXT, CONSTRAINTS, FORMAT, EXAMPLES, OUTPUT.

131

2.   **Avoid Platform-Specific Language**

Don't use "ChatGPT, please…" or emojis unless required. Stay universal.

3.   **Define Output Type and Format Expectations**

Specify what the AI should return: List? Table? Headline? Markdown?

4.   **Use Variable Placeholders**

Write inputs like [TOPIC], [AUDIENCE], or [GOAL] so others can reuse easily.

5.   **Document and Version It**

Create a versioned library or registry with prompt descriptions, authorship, and date.

6.   **Test Transferability**

Try the same prompt in Claude, Perplexity, or GPT and compare consistency. Adjust structure as needed.

---

8. PROMPT TEMPLATE (COPY-READY, MODULAR FORMAT)

ROLE: You are a senior UX copywriter for a mobile-first productivity app.

CONTEXT: The app is launching a new feature that blocks distractions.

TASK: Generate 3 sharp taglines that frame this as a calming, empowering feature — not a control mechanism.

CONSTRAINTS:

- Max 10 words each
- No fear-based language
- Audience: stressed professionals age 25–40

FORMAT: Return as a simple numbered list, no explanation.

EXAMPLES (if needed):

1. Focus Forward, Worry Less
2. Calm In Your Pocket
3. Productivity Meets Peace

---

9. PROMPT OUTPUT EXAMPLE

1. Your Work, Your Focus
2. One Tap to Silence the Noise
3. Built for Flow, Not FOMO

---

10. INDUSTRY FIT & REAL USE CASES

Ideal Industries:

- Product Teams
- Agencies & Consultants
- AI Integration Projects
- Content Libraries & SOP Builders
- Developer-AI Collaboration
- HR, Training, and Onboarding

Real Use Cases:

- Shared prompt libraries across departments
- Client-approved prompt frameworks for agencies
- Reusable onboarding scripts for GPT/Claude agents
- SOPs with embedded prompts for virtual assistants
- Training junior teams with repeatable templates

SYSTEMIC APPLICATION
Use format-for-transfer prompting to create AI agent templates across

133

departments (e.g., marketing, HR, legal) — each with tailored personality and workflow roles.

INDUSTRY FIT BONUS
Operations / Automation: Build one prompt for five teams — each gets a "flavor-adjusted" output with localized voice and compliance rules.

---

11. ADVANCED LAYERS & EXTENSIONS

- **Prompt Registry System:** Track prompt name, use case, last modified, owner
- **Prompt Tags & Metadata:** Add searchable tags: [SALES], [UX], [EMAIL], [CLAUDE-SAFE]
- **Cross-Model Porting:** Translate the prompt into Claude and Gemini-optimized versions
- **Version Control Prompts:** Include v1.0, v1.1 labels to track changes and improvements
- **Auto-Test Scripts:** Build tests that auto-run prompts across multiple AIs and compare

---

12. VALUE SUMMARY

User Level: Beginner
Usefulness: Helps write cleaner, more reusable prompts from day one.

User Level: Intermediate
Usefulness: Enables team collaboration, repeatability, and clarity.

User Level: Expert
Usefulness: Supports prompt scaling, versioning, and multi-model ops.

User Level: CTO / Strategist / Consultant
Usefulness: Future-proofs internal prompt ops and builds prompt-as-IP infrastructure.

User Level: Educator
Usefulness: Great for teaching structured thinking and prompting-as-design.

---

13. LAW SNAPSHOT

Role: Prompt System Designer
Lens: Portability, modularity, transferability
Strength: Clarity, scalability, durability
Weakness: Can feel rigid for creative brainstorming; better suited to repeatable tasks

---

PRO TIPS

Run every new prompt through this law once before saving to a library.
Create a universal prompt format doc for your team — save hours later.
Use Law 1 (Modular Prompt Architecture) to plug this into a larger system.
Teach this format to interns — they'll build smarter, faster.
Create a 'Prompt Passport' — a doc that explains this format to anyone, anywhere.

IDEAS TO TRY

- Use the Cross-Model Transfer Test:

"Write this prompt for ChatGPT, Claude, and Gemini — then analyze what needs to change for each to perform consistently."

Run a Medium Switch Prompt:

"Now take the same idea and reformat it for each of the following:

A 60-second spoken pitch.
A LinkedIn carousel
A technical blog"***
This tests the prompt's logic, not just language.

Create a Format Flex Kit:
For each important prompt in your library, document:

Original structure
Alternate output types (e.g., table, JSON, plain text)
Transfer-ready variations for external use (e.g., SOP, client briefing)

🔗 PROMPT CHAIN LINK

Where this law connects to others. Use together for amplified results.

Law 1: Modular Prompt Architecture — modular prompts are easier to format and move. Combine these laws to build plug-and-play prompt blocks.
Law 17: Recursive Prompt Evolution — update each prompt version with transfer formatting for documentation, reuse, and scaling.
Law 4: Prompt Memory Layering — use formatting to preserve memory across steps when migrating prompts between chains.
Law 16: Multi-AI Role Rotation — when rotating roles across

models, transfer formatting ensures consistency in tone and output structure.

Unlike Law 1, which builds modular prompts, Law 15 is about making your prompts transferable between models or tools. It's the packaging layer — not the blueprint.

---

14. LAW 15: FORMAT-FOR-TRANSFER PROMPTING [TRAIN IT TO TRAVEL] FINAL TAKEAWAY

A smart prompt isn't just brilliant once — it's usable again, anywhere, by anyone.
Train your prompts like systems — not sentences.

LAW 15 BONUS – PROMPT-TO-SYSTEM UPGRADE

"Prompts are single shots. Systems are symphonies."

Insightful Tip:
Build reusable templates. Name them. Save them. Use them like custom GPTs — even without coding.

Real-World Prompt Flip:

"Create a system called 'PromptPressKit' for writing thought leadership posts. Include 3 reusable prompt slots."

Mini-Mindset Shift:
If it worked once, don't recreate it.
Systemize it.

LAW 16: Multi-AI Role Rotation [Cross-Model Thinking]

Different AIs think differently — rotate your bench, not just your prompt.

---

1. LAW SUMMARY

This law teaches how to leverage multiple AI models by assigning them distinct roles or perspectives, rather than running the same prompt across each one. You rotate their mental models — not just their names — and use their differences as creative and critical force multipliers.

---

2. METAPHORS / QUANTUM SLOGANS

- *Don't just use multiple AIs — give them different minds.*
- *Claude is your philosopher. GPT is your strategist. Gemini is your analyst.*
- *Smart prompting isn't multitasking. It's multi-model thinking.*
- *Why settle for one brain when you can build a boardroom?*
- *Don't prompt in parallel. Rotate and orchestrate.*

---

3. MINI MANIFESTO / HOOK

Most people try GPT. Then maybe Claude. Maybe Gemini next.

They run the same prompt... and compare output.
That's fine for checking quality — but not for building intelligence.

This law flips the strategy:
Assign each AI a unique role, lens, or mindset.
You turn one task into a panel of advisors — each with a different specialty, bias, or background. Then you compare, extract, and **synthesize real insight**.

Stop switching models like browsers.
Start rotating them like minds.

---

4. FOUNDATIONAL IDEA (HUMAN SIDE + AI SIDE)

Human Side:
Cognitive diversity drives better decision-making. It's why the best teams include engineers, artists, skeptics, and dreamers. This law applies the same principle — using different AIs as different roles in your mental think tank.

AI Side:
Each model has a different "personality" based on training data, tone calibration, reasoning preferences, and risk tolerance. Instead of ignoring that — this law makes it a strategy.

---

5. CORE IDEA (EXECUTIVE SUMMARY)

139

Multi-AI Role Rotation means using several models — GPT-4, Claude, Gemini, and others — but assigning **different prompt roles** to each:

- One as the optimistic marketer
- One as the skeptical critic
- One as the legal/compliance officer
- One as the product manager
- One as the Gen Z buyer

Then you gather, analyze, and synthesize their responses into something smarter than any one model alone.

---

6. WHAT IT IS (FOUNDATIONAL CONCEPT)

Instead of redundancy (same prompt, different models), this law prompts for **cognitive variance**:

- Rotate tone
- Rotate worldview
- Rotate function (ideator, editor, critic, historian)
- Rotate risk profile (safe vs. bold)

You're not just testing outputs. You're creating **AI triangulation** — using difference to find signal.

---

7. HOW TO USE IT (STEP-BY-STEP)

1. **Define the Task or Message**
What needs multi-perspective input?

2. **Assign Roles to Each Model**

Decide each AI's persona: visionary, devil's advocate, investor, compliance officer, etc.

3. **Tailor the Prompt Per Role**

Each model gets a slightly different angle — same core task, different mindset.

4. **Compare Responses**

Where do they converge? Where do they disagree?

5. **Synthesize or Stack**

Build the final answer, product, or strategy from the *best parts* of each.

---

8. PROMPT TEMPLATE (COPY-READY, MODULAR FORMAT)

TASK: We're launching a smart fitness wearable for Gen Z.

Model 1 (GPT-4): You are the brand strategist. Write the value proposition and product hook.

Model 2 (Claude): You are an ethical psychologist. Highlight potential emotional risks in the messaging.

Model 3 (Gemini): You are a digital trends analyst. Predict adoption friction points based on recent Gen Z behavior.

FINAL: Based on all 3, synthesize a go-to-market message that's compelling, ethical, and trend-aligned.

---

9. PROMPT OUTPUT EXAMPLE

GPT-4 (Strategist): "Train smarter, not harder — with a wearable that adapts to your body and your goals."

Claude (Ethics): "Avoid framing around insecurity or perfectionism. Focus on empowerment and autonomy."

Gemini (Trends): "Gen Z is skeptical of 'tracking everything.' Emphasize opt-in data control, not surveillance."

Final Synthesis Prompt (back to GPT-4):
"Now write a product headline and subline that reflects this synthesis: smart adaptation, emotional empowerment, and opt-in control."

10. INDUSTRY FIT & REAL USE CASES

Ideal Industries:

- Product Development
- Strategic Messaging
- Startup Pitches
- Education & EdTech
- Policy & Ethics
- AI Integration Teams

Real Use Cases:

- Launching new tech products
- Reviewing marketing for emotional tone or generational fit
- Testing political framing across ideological lenses
- Comparing AI-generated SOPs for tone and bias
- Training teams in critical synthesis via AI panels

---

11. ADVANCED LAYERS & EXTENSIONS

- **Model Mapping Matrix:** Build a spreadsheet of model strengths (e.g. Claude = nuance, Gemini = speed, GPT = logic/creativity balance)
- **Cross-Echo Comparison:** Prompt all models to critique the same message from their assigned roles

- **Simulation Chain:** Have each AI rewrite the other's response from its own lens (e.g. Claude refines GPT's marketing copy for ethical tone)
- **Agent Orchestration:** Use LangChain, CrewAI, or manual methods to run these roles in flow
- **Future Layering:** Rotate models based on task phase — ideate (Gemini), refine (Claude), package (GPT)

TEST CASE METHOD
Feed the AI three prompts:

- One good
- One flawed
- One dangerous
  Ask it to compare them without knowing which is which.
  You'll uncover its blind spots — and yours.

---

12. VALUE SUMMARY

User Level: Beginner
Usefulness: Introduces AI diversity — helps pick the best model for the task.
User Level: Intermediate
Usefulness: Trains lateral thinking, multi-angle analysis, and prompt personalization.
User Level: Expert
Usefulness: Enables cross-model synthesis for higher-quality, bias-resistant outputs.
User Level: CTO / Strategist / Consultant
Usefulness: Creates a future-proof AI stack that maximizes model variance and insight coverage.
User Level: Educator

Usefulness: Teaches critical thinking, model fluency, and collaborative reasoning.

---

13. LAW SNAPSHOT

Role: AI Orchestrator / Perspective Rotator
Lens: Multi-model logic, contrast synthesis, cognitive diversity
Strength: Innovation, critical foresight, bias resistance
Weakness: Can become overwhelming without structured comparison or time limits

---

PRO TIPS

• *Law 16 rotates between viewpoints — your user, your critic, your competitor. It's broader than just assigning roles (Law 2) or self-feedback (Law 6). It's a cognitive expansion tool.*
• *Don't just rotate models — rotate personas. That's where the depth comes from.*
• *Build a "model character guide" for your team to remember each AI's voice.*
• *Use Law 9 (Mirror Self Protocol) inside each model to simulate "you in that role."*
• *Turn this into a workshop or training simulation — great for upskilling teams in AI collaboration.*
• Assign models not just by strength — but by tone. Use GPT for synthesis, Claude for empathy, Gemini for data.

IDEAS TO TRY

- Use the **AI Council Method**:

"Claude is the cautious lawyer. GPT is the visionary CMO. Gemini is the logical COO. Here's the challenge — debate it."
Synthesize the outcome across lenses.

- Run a **Role Rotation Roundtable**:

Round 1: GPT acts as advisor
Round 2: Gemini challenges GPT
Round 3: Claude offers the neutral summary
Then combine all three into a master insight set.

- Create a **Model Personality Map**:
Document which models excel at what:

Claude = *nuance, ethics, structure*
GPT = *persuasion, strategy, storytelling*
Gemini = *logic, brevity, planning*
Use this to assign each prompt to its ideal thinker.

PROMPT CHAIN LINK

Where this law connects to others. Use together for amplified results.

Law 9: Mirror Self Protocol — simulate future-you across different AIs and see which one reflects your best reasoning.
Law 6: Cognitive Clone Critique — use multiple models to simulate different versions of "you" or your team personas.
Law 14: Ethical Duality Simulation — assign conflicting ethical frames to different AIs to see which moral narrative holds strongest.
Law 15: Format-for-Transfer Prompting — ensure your prompts maintain structure and tone when rotated between models and use

cases.

Use with **Law 11: Ethical Duality Simulation** and **Law 6: Cognitive Clone Critique** to simulate disagreement and audit hallucinations in early drafts.

---

14. LAW 16: MULTI-AI ROLE ROTATION [CROSS-MODEL THINKING] FINAL TAKEAWAY

The smartest answer isn't the one from the smartest model.
It's the one you built by orchestrating many minds — not just switching them.

LAW 16 BONUS – PERSONA STRETCHING & THINKING TACTICS

"AI isn't just a tool. It's a lens. The persona you assign is the mind you borrow."

Insightful Tip: When you change the role you assign to AI, you also change the lens through which you think. Prompting multiple personas helps you:

Break mental echo chambers

Reveal hidden assumptions

Imagine from angles you normally ignore

Real-World Prompt Flip: "Now respond as: 1) a skeptical investor, 2) a Zen monk, and 3) a visionary 2030 futurist. What would each advise about this plan?"

This prompt doesn't just test content — it expands perspective. It builds mental agility and cognitive range.

Mini-Mindset Shift: You don't just prompt the AI.
You prompt yourself through AI — from every angle that matters.

LAW 17: Recursive Prompt Evolution [Build the Next Version]

The best prompt isn't the one you write. It's the one you evolve.

---

1. LAW SUMMARY

*This law teaches you how to turn prompting into a recursive system — where every output becomes a signal for the next, every mistake becomes training data, and your prompt library becomes smarter with use. You stop prompting once. You start **evolving forever**.*

---

2. METAPHORS / QUANTUM SLOGANS

- *Prompting isn't an event. It's a loop.*
- *Every prompt is a prototype. Every output is feedback.*
- *Build the prompt that builds better prompts.*
- *Don't aim for perfect. Aim for upgradeable.*
- *Your prompt isn't done when it works — it's done when it grows without you.*

---

3. MINI MANIFESTO / HOOK

Most users stop when a prompt works.
Power users **start** when it works — because now they can improve it.

This law unlocks your final shift:
From one-time prompting → to **recursive prompting systems**.
You don't just ask for content. You ask AI to evaluate itself, revise its own prompt, test alternate versions, and suggest upgrades.

It's not just productivity.
It's **prompt evolution** — the bridge between human strategy and AI self-improvement.

---

4. FOUNDATIONAL IDEA (HUMAN SIDE + AI SIDE)

Human Side:
We learn by doing, reflecting, and iterating. AI prompting can — and should — mirror that. You build feedback loops into the workflow so prompts get stronger, faster, and more resilient.

AI Side:
LLMs can critique, compare, and revise prompts — especially if you ask clearly. Recursive prompting allows the AI to:

- Suggest improvements to your own prompt
- Refine outputs with targeted instructions
- Create its own meta-prompts for better performance

This turns the AI into its own **editor, tester, and trainer.**

---

5. CORE IDEA (EXECUTIVE SUMMARY)

Recursive Prompt Evolution means treating every prompt as a **living system**. Instead of stopping at "good enough," you build loops:

- *Evaluate the output*
- *Refine the prompt*
- *Re-run the improved version*
- *Repeat with smarter constraints or role shifts*

This technique is especially valuable when:

- Building prompt libraries
- Running long-term AI workflows
- Training new team members on effective prompting
- Optimizing high-stakes or high-volume outputs

---

6. WHAT IT IS (FOUNDATIONAL CONCEPT)

This law uses AI for **meta-prompting** — asking it to:

- Critique the current prompt's performance
- Suggest alternate phrasings or structure
- Improve clarity, logic, or modularity
- Explain why one version might outperform another
- Log strengths, flaws, and upgrade history

It's prompt engineering — **that evolves itself.**

---

7. HOW TO USE IT (STEP-BY-STEP)

1. **Write and Run the Initial Prompt**
Test it on your task, get output, and identify gaps.
2. **Ask AI to Review Its Own Output**
"What's weak or unclear in this result?"

3.   **Ask AI to Suggest Prompt Edits**
"How would you rewrite the prompt to fix those issues?"
4.   **Compare Old vs. New Performance**
Run both prompts. Ask AI to judge which works better — and why.
5.   **Log, Label, and Save the Upgraded Version**
Track versions, date, and author. Use comments to preserve lessons learned.

---

8. PROMPT TEMPLATE (COPY-READY, MODULAR FORMAT)

ORIGINAL PROMPT: [Paste prompt]
TASK:

1.   Evaluate this prompt's effectiveness on clarity, modularity, and performance.
2.   Suggest a rewritten version that might perform better.
3.   Briefly explain why the change would improve the output.
4.   (Optional) Run both prompts on the same task and compare.

---

9. PROMPT OUTPUT EXAMPLE

Evaluation:

- Lacks tone direction
- Output format unclear
- Includes nested instructions that may confuse the model

Suggested Rewrite:
ROLE: You are a B2B product strategist.
TASK: Generate 5 landing page headlines for a cloud security

platform.
CONSTRAINTS:

- Max 8 words
- Tone: Urgent but not fear-driven

FORMAT: Numbered list, no explanation

Why it's better: Cleaner task structure, more specific constraints, and clearer tone guidance.

---

10. INDUSTRY FIT & REAL USE CASES

Ideal Industries:

- SaaS Teams Building AI SOPs
- PromptOps for Enterprises
- Research and Think Tanks
- UX Writing and Personalization
- Knowledge Management & L&D

Real Use Cases:

- Building adaptive prompt libraries
- Improving long-form workflows like proposal writing
- Training teams in prompt best practices
- Auto-improving AI onboarding assistants
- Creating "prompt QA" systems in high-stakes settings

---

11. ADVANCED LAYERS & EXTENSIONS

152

- **Prompt Scoring System:** Ask AI to rank prompt performance 1–10 across multiple criteria
- **Compare-and-Evolve Chain:** Run v1, v2, and v3 — and evolve each generation
- **Self-Improving Prompt:** "Every time you run this, suggest how to improve it next time"
- **Automated Logging:** Use AI to create a changelog after each prompt evolution
- **Team Prompt Feedback Loop:** Combine Law 9 (Mirror Self Protocol) with recursive logic for peer-style review

---

12. VALUE SUMMARY

User Level: Beginner
Usefulness: Reinforces the idea that prompts can — and should — improve over time.

User Level: Intermediate
Usefulness: Helps detect weaknesses, revise prompts, and train others with evolving best practices.

User Level: Expert
Usefulness: Enables scalable prompt design systems with built-in QA and learning feedback loops.

User Level: CTO / Strategist / Consultant
Usefulness: Builds future-proof prompt infrastructures that get smarter with every use — turning IP into a living system.

User Level: Educator
Usefulness: Teaches iterative reasoning, self-reflection, and long-term optimization — core skills for the AI-powered workforce.

13. LAW SNAPSHOT

Role: Prompt Evolution Architect
Lens: Iteration, QA, self-refinement

Strength: Longevity, clarity, system growth
Weakness: Can lead to overengineering or "infinite tweaking" if no stopping rule is set

---

PRO TIPS

Every strong output is a seed for a better prompt. Recursive prompting turns good into great — by design, not by luck.

Don't stop at success. Run a second prompt: "Rewrite the original to improve clarity, precision, and adaptability — based on this result."

This law transforms AI from a static tool into an evolving creative partner.

Ask the AI: "What assumptions is this prompt silently making?" Then ask: "If those assumptions are wrong — what changes?"

WARNING
Don't use hallucination detection prompts as **final output testers**. They're for diagnostics — not for publishing safety.

Some users misunderstand this Law as a fact-checker instead of a hallucination lens.

IDEAS TO TRY

Use the Prompt-as-Code Loop: *"This prompt worked. Now debug it like software:*
What variable is hardcoded?

What constraint is unclear?
How can it be modularized?" *

Run a 2-Level Self-Improvement Chain:

1.   Prompt A produces Output A
2.   Prompt B asks AI: *"Refactor Prompt A to generate Output A more clearly and scalably."*
3.   Test new Prompt A2 against A.
Repeat weekly for evolving libraries.

Document Prompt Version History:
Use tags like:
V1.0 — initial test
V1.1 — optimized tone
V2.0 — modular and reusable
Treat prompts like assets — with version control and naming conventions.

PROMPT CHAIN LINK

Where this law connects to others. Use together for amplified results.
Law 1: Modular Prompt Architecture — recursive prompting evolves each module independently for clarity and reuse.
Law 3: Prompt Autopsy & Debugging — feed failed prompts into this law to produce refined, stronger versions.
Law 15: Format-for-Transfer — build prompts that improve not just in logic, but in adaptability for multi-channel use.
Law 16: Multi-AI Role Rotation — evolve prompt versions using feedback from different models to create hybrid best-practice formats.

Law 17 builds on the modularity from Law 1 but focuses on evolution: improving prompts over time using recursive feedback. It's about growing a smart prompt, not just building or reusing it.

14. LAW 17: RECURSIVE PROMPT EVOLUTION [BUILD THE NEXT VERSION] FINAL TAKEAWAY

AI doesn't just respond. It learns — if you let it.
And the smartest prompters aren't writers. They're builders of better versions.

LAW 17 BONUS – AI-ASSISTED META-THINKING

"AI can't replace your thinking. But it can make your thinking visible."

Insightful Tip:
Use prompts to map your logic, challenge your biases, or stress-test decisions. It turns invisible hunches into editable maps.

Real-World Prompt Flip:

"Break my logic into 5 steps. Then challenge step 3 with a counterexample."

Mini-Mindset Shift:
Prompting isn't just output.
It's outsourced introspection.

LAW 18: Prompt Portfolio Intelligence [Prompt-as-a-Service Method]

Don't just prompt — invest.
 Build a system where your best prompts grow, evolve, and earn like assets.

---

1. LAW SUMMARY

Every brilliant prompt deserves a second life. Turn your most effective prompts into reusable frameworks, evolving tools, and strategic assets. This law teaches you to think like a prompt investor—not just a prompt user.

---

2. METAPHORS

Prompts as Software: Code once, deploy many times.
Prompt Vault: A creative bank account that earns compound interest.

---

3. MINI MANIFESTO / HOOK

Most users lose 90% of their best prompts. The smartest ones don't let them vanish.

They archive, refine, and scale them. This law elevates prompting from ephemeral output to permanent, modular asset.

157

With Prompt Portfolio Intelligence, every session becomes a step toward a system

---

4. FOUNDATIONAL IDEA (HUMAN SIDE + AI SIDE)

Human Side:
Like great thinkers who keep notebooks or engineers who build libraries, strategic prompt creators treat effective prompts as reusable components. This draws from knowledge management systems, design thinking, and IP documentation.

AI Side:
LLMs don't just react to prompts—they expose underlying logic patterns. When you treat prompts as evolving blueprints with structure, tags, and modularity, the AI adapts faster, recalls better, and outputs smarter content across tools and models.

---

5. CORE IDEA (Executive Summary)

Prompt Portfolio Intelligence is the shift from prompting reactively to managing prompts like a library of strategic intellectual property. Just as engineers don't rewrite every line of code from scratch, expert AI users shouldn't start over with every new query. Instead, they store, refine, and scale prompts across contexts, AIs, and industries.

B nWith this mindset, you go beyond one-off prompting and begin building a **Prompt-as-a-Service** (PaaS) operation. Whether for yourself, a team, or clients—this law enables versioning, scaling, and monetizing your best thinking.

---

6. WHAT IT IS (Foundational Concept)

Most users lose 90% of their best prompts. This law draws from:

- **Knowledge Management Systems (KMS)**
- **Design Pattern Libraries**
- **SaaS Productization Strategy**

Your prompt becomes a reusable tool:

- With variables
- Use-case tagging
- Feedback notes
- A/B versions
- Licensing potential

This is how prompt engineering evolves into prompt operations.

---

7. HOW TO USE IT (Step-by-Step)

1. **Capture** every successful prompt you run (don't trust memory).
2. **Label** it clearly: intent, tone, model used, ideal context.
3. **Tag** the prompt with searchable categories (e.g. [SALES], [UX], [CLAUDE-SAFE]).
4. **Version** your improvements: 1.0, 1.1, 2.0.
5. **Test Transferability**: Run on multiple AIs (GPT, Claude, Gemini).
6. **Document Output** differences and required adaptations.
7. **Share or Package**: Offer as templates, downloads, or paid assets.

---

8. PROMPT TEMPLATE (COPY-READY)

ROLE: *You are an AI Prompt Portfolio Architect.*

TASK: *Convert the following high-performing prompt into a reusable asset.*

INPUT PROMPT: [Insert existing prompt here.]
OUTPUT FORMAT:

- Prompt Title:

- Ideal Use Case:

- Key Variables:

- Model Notes:

- Version:

- Sample Output:

- Tags:

---

9. PROMPT OUTPUT EXAMPLE

Prompt Title: *Product Naming – Emotional Benefits Focus*
Ideal Use Case*: Naming digital products in health & wellness*
Key Variables*: Audience = Gen Z, Emotion = Empowerment, Limit = under 5 words*
Model Notes*: Works best with GPT-4 and Claude 3*

Version: *v2.1*

160

Sample Output:

1. Calm Within

2. MindLift

3. Feel Forward

4. InnerFuel

5. ThriveCore

Tags: *[Naming], [Emotion-Driven], [Wellness], [Short]*

---

10. INDUSTRY FIT + REAL USE CASES

Ideal Industries:

- Marketing & Copywriting
- SaaS & Product Design
- Education & Coaching
- Legal & Compliance

Real Use Cases:

- Agencies reusing prompts across client verticals
- Startups training AI agents with prompt libraries
- Teachers creating lesson-generating templates
- Legal teams documenting safe prompt versions

---

11. ADVANCED LAYERS & EXTENSIONS

Prompt Registry Systems: Shared team libraries with access levels

Cross-AI Optimization: Prompts optimized per model family

Auto-Evolution Scripts: AI that improves and versions its own prompts

Prompt Monetization Models: Licensing or selling use-specific packs (Gumroad, Patreon, Notion)

VISUAL REFERENCE IDEA
Add a tag to each saved prompt:
"Fails if..." — followed by an emoji, traffic light, or risk label. Build a prompt library that not only shows success but warns of misuse.

12. VALUE SUMMARY (USER LEVEL MATRIX)

User Level: Beginner
Usefulness: Encourages saving prompts early and intentionally.
User Level: Intermediate
Usefulness: Enables repeatable workflows and faster iterations.
User Level: Expert
Usefulness: Builds structured IP, improves cross-project intelligence.
User Level: CTO / Strategist / Consultant
Usefulness: Supports scalable prompt design, documentation, training.
User Level: Educator / Trainer
Usefulness: Creates teachable prompt logic and thinking modules.

13. LAW SNAPSHOT

Role: Prompt System Designer / Creative Ops
Lens: Modularity, reusability, intellectual capital
Strength: Operationalizes high-performing prompt logic
Weakness: May become bloated without clean tagging/archiving

PRO TIPS

- Create a shared doc or Notion page titled: "Prompt Library v1.0"
- Run Law 1 (Modular Prompt Architecture) inside every saved prompt
- Schedule quarterly audits of your top-used prompt assets
- Add sample outputs next to each prompt
- Create a 1-page visual map linking related prompt packs

IDEAS TO TRY

Use **Prompt Cloning**: Replicate prompt logic across industries (e.g., use same structure for "coach", "therapist", "UX writer")
Run **AI Audit Reviews**: Ask Claude or GPT to evaluate your prompt library
Build a **Prompt Marketplace Trial**: Sell one pack on Gumroad as a test.
Try the **Model Switch Test**: Run same prompt on GPT, Claude, Gemini, and Midjourney.
Launch a "Prompt-as-a-Newsletter" where readers get one strategic prompt per week

---

CROSS-LINKS (USE TOGETHER)

Law 1: Modular Prompt Architecture — Use modularization to build better prompt components.

Law 3: Prompt Autopsy Method — Version your prompts by learning from what fails.
Law 6: Cognitive Clone Critique — Use AI to self-critique your prompt system.
Law 16: Multi-AI Role Rotation
Law 17: Recursive Prompt Evolution

EXPERT-LEVEL APPLICATION

Build a **Prompt Ops Handbook** for your organization. Include:

- Naming conventions
- Format standards
- Ethical tags
- Approved model compatibility
- Output benchmarks

Make it part of onboarding and team rituals. This turns prompt design into an org-wide skill.

---

14. LAW 18: PROMPT PORTFOLIO INTELLIGENCE FINAL TAKEAWAY

A great prompt isn't a moment. It's a system. Build portfolios, not just sessions. Version, evolve, and deploy your thinking.

LAW 18 BONUS – PROMPT ASSETS, NOT PROMPT MOMENTS

"A saved prompt is a future product. A lost prompt is wasted capital."

Insightful Tip:

Start a personal Prompt Bank using Notion, Obsidian, or Google Sheets.
Tag each prompt with: context, model, variables, and a "best use" snapshot.
Bonus: Add a risk flag — how it might fail or hallucinate.

Real-World Prompt Flip:

"Take my last 5 best-performing prompts and version them into reusable
frameworks. Add tags, variables, and model notes."

Mini-Mindset Shift:

Prompts are not throwaways. They're prototypes of intelligence.
Treat them like modular code — and you'll build faster, smarter, and more
strategically than anyone stuck in "prompt > response > forget."

LAW 19: Multimodal Prompting Mastery

Prompt beyond words. *Shape intelligence across vision, sound, and space — not just language.* ***Prompting Across Text, Image, Audio, Video, and Multi-Agent AI***

---

1. LAW SUMMARY

Multimodal prompting is the art of speaking to AI in multiple formats — image, text, voice, video, and beyond. This law unlocks next-gen creative collaboration by teaching you how to strategically combine formats to achieve richer, more intelligent outcomes.

---

2. METAPHORS / QUANTUM SLOGANS

- Don't just write prompts — *paint*, *perform*, and *direct* them.
- Multimodal prompting is the cinematic directing of AI.
- One sense limits the output. Many senses unlock intelligence.
- Stop prompting like it's radio. Start prompting like it's IMAX.

---

3. MINI MANIFESTO / HOOK

Text prompting is powerful — but it's only one instrument in the orchestra.
Vision, audio, video, and interaction create a symphony of

intelligence. The more senses AI can perceive, the smarter — and more useful — it becomes.

4. FOUNDATIONAL IDEA (HUMAN + AI SIDE)

Human Side: Humans communicate in layers: speech, gestures, visuals, tone. Limiting your AI to text-only prompts is like running a company with one department. Multimodal prompts let you activate more cognitive power — and more creativity.

AI Side: Multimodal models like GPT-4V, Claude Sonnet, Gemini Pro, and Sora can process not only text, but images, charts, audio clips, and videos. Each mode activates a different pattern-matching layer. Combining them increases reasoning, retention, and realism.

5. CORE IDEA (EXECUTIVE SUMMARY)

Multimodal Prompting Mastery teaches you how to think beyond words — and design inputs that mirror how humans perceive.

This includes:

- Prompting images and screenshots (e.g., UI mockups, charts)
- Using video context to guide narrative generation
- Summarizing or remixing YouTube videos via transcript + tone cues
- Adding voice prompts or tonal emotion
- Blending formats to create richer, cross-contextual AI responses

6. WHAT IT IS (FOUNDATIONAL CONCEPT)

Multimodal Prompting means:

- **Image-to-text prompting** (e.g., "describe this graph")
- **Text-to-image prompting** (e.g., Midjourney, DALL·E)
- **Video summarization prompts** (via URL + task + voice tone)
- **Audio transformation prompts** (convert voice → blog, blog → voice)
- **Cross-modal fusion** (feed image + prompt + persona together)

---

7. HOW TO USE IT (STEP-BY-STEP)

1. **Choose your target output** (e.g., image, voiceover, storyboard)
2. **Select your input mix** (e.g., text + video, or voice + transcript)
3. **Use sensory-specific framing:**

- For image: "What's happening in this photo — as a UX designer would see it?"
- For audio: "Change this podcast clip into a CTA in 10 seconds."

4. **Clarify your constraints** (e.g., style = cinematic, tone = optimistic)
5. **Add context in layers** (audience, persona, backstory)
6. **Preview and test output across tools**
7. **Refine using chain prompting or visual iteration**

MULTIMODAL CONTRAST METHOD

Feed the same visual to GPT, Claude, and Gemini.
Then prompt each AI to explain what *another model might miss*.
This builds diagnostic intelligence — not just content generation.

168

INDUSTRY FIT BONUS
Customer Support: Feed transcripts + tone models into AI to build multilingual, emotionally aware chatbots that adapt tone based on input medium (email vs. voice).

8. PROMPT TEMPLATE (MODULAR)

INPUT: *[Paste video link or image + your written context]*

TASK: *Summarize or redesign this content for [Audience] in [Format] using [Tone/Style]. Highlight [Key Message or Emotion].*

CONSTRAINTS: *Keep it under [Length] and make it useful for [Platform].*

---

9. PROMPT OUTPUT EXAMPLE

INPUT: Screenshot of an outdated website homepage
TASK: Redesign this as a modern SaaS landing page
OUTPUT: Clean, minimal layout with bold headline, central CTA, testimonials under the fold, muted color palette, tech-oriented font.

---

10. INDUSTRY FIT + REAL USE CASES

- **Marketing**: Turn blog posts into visuals, or whitepapers into videos
- **Education**: Turn lecture notes into narrated animations
- **Healthcare**: Summarize medical image reports or patient videos
- **Legal**: Convert court transcripts into digestible case timelines
- **Design**: Test reactions to visual prototypes with AI-generated voiceovers

169

11. ADVANCED LAYERS & EXTENSIONS

- Prompt Chaining Across Formats (text → image → CTA → video)
- Vision Critique Loops: AI critiques its own visual output
- Audio Emotion Layers: Detect tone shifts and revise accordingly
- Caption-Driven Style Control: Use image captions to steer AI style
- Prompt Remixing: Same content, different medium (e.g., voice vs. carousel)

12. VALUE SUMMARY (USER LEVEL MATRIX)

Beginner: Converts blog to image, or LinkedIn post to video
Intermediate: Combines formats to create richer content
Expert: Designs cross-modal learning systems or agent behaviors
Consultant: Builds multimodal AI pipelines across brands and tools
Educator: Teaches cross-format storytelling and engagement

13. LAW SNAPSHOT

Role: Multimodal Prompt Designer
Lens: Multi-sensory communication + human-grade intelligence
Strength: Realism, engagement, cross-AI performance
Weakness: Requires tool access and sensory-specific skill sets

PRO TIPS

Use image prompts as "setup context" before issuing task commands.
Feed AI your visual content before writing — it changes the logic order.
Use Descript + ElevenLabs to turn high-performing prompts into voiceover videos.
For image tools (e.g., Midjourney), write text prompts *as if describing a movie scene*

IDEAS TO TRY

- Try prompting GPT-4V to rewrite your sales page *after* analyzing its layout screenshot
- Feed Claude a transcript + your video style — then ask it to rewrite your YouTube script in that tone
- Generate 3 variations of a Midjourney image, then have GPT analyze which fits your brand better

---

CROSS-LINKS (USE TOGETHER)

Law 1: Modular Prompt Architecture — Create swappable input/output format modules.
Law 15: Format-for-Transfer Prompting — Translate prompts across formats and tools.
Law 17: Recursive Prompt Evolution — Improve multimodal prompts over time

---

14. LAW 19: MULTIMODAL PROMPTING MASTERY – FINAL TAKEAWAY

The smartest prompts aren't just written — they're seen, heard, and felt.
Multimodal prompting is how you speak to the full intelligence of the system.

Language is just one lens.
True prompting happens when your ideas speak, move, and see.
Multimodal isn't a feature — it's the evolution of how we instruct intelligence.

LAW 19 BONUS – SENSE-WIDE STRATEGY

"Prompting is no longer just language. It's choreography."

Insightful Tip:

Use multimodal prompts to test idea fidelity across modalities.
Start with one core idea, then express it as:

- A text headline (GPT)
- A voiceover (ElevenLabs)
- An image prompt (DALL·E or Midjourney)
- A video sketch (Runway or Sora)

Compare: What's preserved? What's lost?

Real-World Prompt Flip:

"Take this single idea: [insert idea].
Now:

1. Prompt GPT-4 for a slogan
2. Prompt Gemini for supporting research
3. Prompt DALL·E for a symbolic image
4. Prompt ElevenLabs to narrate the message
5. Prompt Sora (or Runway) for a teaser video idea."

172

Bonus Application: Using This Book Itself as a Multimodal Prompt Case

Want to prove you understand multimodal prompting?
Use this book as your playground. Here's how:

MidJourney Prompt (Visual Metaphor for Book Marketing):

/imagine
"A futuristic library where AI avatars are studying glowing scrolls titled 'The 19 Laws of Prompting.' In the center: a human in meditation pose surrounded by floating prompts encoded like constellations."
--ar 16:9 --style raw --v 6 --q 2

This isn't just a promo image — it's visual storytelling of the book's mission.

Sora Prompt (Short Film or Trailer Concept):

Prompt:
"Create a 15-second cinematic teaser for a nonfiction book about AI prompting mastery. Show a user whispering a prompt into thin air — and watching it ripple into blueprints, headlines, conversations, videos, and art. End with the phrase: 'Prompt once. Influence everything.'"

Bonus twist: Ask Sora to generate **multiple camera perspectives** — one from the human's eye, one from the AI's POV, and one as a third-person flyover.

Meta-Prompting Tips:

After generating both: Ask GPT - "What do these outputs fail to convey that the book explains better?"
Then reverse-engineer that insight into your next prompt.

Don't prompt AI. Direct intelligence.
You're not a writer. You're a cross-sensory strategist.
Use every channel — not to impress the AI, but to align human experience.

Sora Prompt: AI Book Teaser (15 sec)

Prompt: "A sleek AI lab blends with a classical library. A human touches a glowing prompt floating in the air — it bursts into motion: one becomes many. Visuals of AI writing headlines, generating images, voice scripts, and simulating debates unfold rapidly. In the background, glowing text reads: 'LAW 1... LAW 7... LAW 19...'
Final scene: the book cover lands softly on a desk as the title appears:
'The 19 Laws of AI Prompting Intelligence – by Nikolay Gul'
Overlay voice or text: **'Prompt once. Influence everything.'**"

Style & Technical Additions (for Sora prompt engineering):

- Cinematic lighting, modern-futuristic environment
- Clean interface overlays (like AR, not neon)
- Soft music pulse or sci-fi piano
- Subtitle-friendly spacing and clarity
- Final 3 seconds must pause on the book title
- 16:9 aspect ratio, crisp depth-of-field effects

Optional Prompt Companion Line (for ChatGPT to enhance with Sora):

"Now make it feel like the viewer just unlocked the user manual for thinking in the future."

Bonus Chapter: Frameworks Behind the Laws – CAB & The AI Decision Compass (YIMBS)

How Two Simple Tools Created Every Law in This Book

"The best laws aren't written first — they're lived first."

Every law in this book wasn't just imagined.
It was pressure-tested, revised, and co-engineered through breakdowns, breakthroughs, and two powerful tools that exposed flaws before they ever reached you.

FRAMEWORK #1: Cross-AI Brainstorming (CAB)

"One AI is clever. Many AIs with clashing opinions? That's intelligence."

Relevant Laws:
Law 10 – Role Simulation Prompting
Law 11 – Echo Chamber Breaker Blueprint
Law 13 – Robustness Testing via Adversarial Framing
Law 16 – Multi-AI Role Rotation

HOW IT WORKS:
Assign different AI agents conflicting roles — The Critic, The Dreamer, The CEO, The Skeptic, The Customer.
Give them the same challenge.
Let them debate. Let them disagree.
Then synthesize their tension into something no single model could offer alone.

PRO TIP:
CAB doesn't just generate better answers — it exposes your blind spots.
It lets you think with friction, not fantasy.

FRAMEWORK #2: The AI Decision Compass

A strategic self-simulation tool for clearer, bias-resistant decisions

"Don't ask AI to flatter you. Ask it to outsmart you — kindly."

Relevant Laws:
• Law 3 – Prompt Autopsy Method
• Law 7 – Bias Mapping Layer
• Law 15 – Format-for-Transfer Prompting
• Law 17 – Recursive Prompt Evolution

How It Works:
Instead of asking AI to please you, prompt it like this:

"Pretend you are me — but smarter, more honest, and allergic to BS. What would you fix, challenge, or rethink in my logic?"

The **AI Decision Compass** became our internal north star.
It shredded motivational fluff.
It questioned beautiful nonsense.
It forced precision.
It made thinking honest again.

Real Result:

An early title for this book was "The AI Genius Framework."
The AI Decision Compass killed it with one line:

"You're feeding ego bias. This book is about collaboration — not worship."

It was brutally right.
And that's exactly the kind of voice this framework gives you:
Detached. Honest. Future-minded.

"CAB cracks you open.
The AI Decision Compass keeps you from sealing shut again."

Real-Life Prompt Flip

Use CAB to test your next business idea with four roles:
Optimist, Pessimist, Operator, Customer.

Then run the AI Decision Compass:

"What part of this plan still feels vague, weak, or self-congratulatory?"

FINAL TAKEAWAY

CAB shows you where you're hiding from the truth.

The AI Decision Compass shows you who you could be if you faced it.

Together, they help you:

• Test smarter product ideas and pitch decks

• Run AI prompting loops that actually learn

• Debug your thinking before public failure

• Replace delusion with durable direction

These aren't just AI tools.
They're cognitive upgrades - for anyone building the future.

Quick Prompts Index – Use These Anywhere

Decision Compass Prompt: "Simulate a version of me who's smarter, honest, and detached from ego. What would they do?"

Prompt Autopsy (Law 3): "Analyze my prompt in 3 parts: what's clear, what's vague, and what's leading the model astray."

Cross-AI Brainstorming: "Have four personas debate this idea: Optimist, Skeptic, Operator, and First-Time User."

Ethics Tension Test: "Flip the moral position. What would an opposing value system say is wrong here?"

Format-for-Transfer Prompting: "Rephrase this so a teammate with no context can use it perfectly."

Silence Amplification (Law 12): "What's NOT being said — and what might that reveal?"

Troubleshooting AI Responses – Find the Fix Fast

Pro Tip: *If you're solving 2–3 problems at once?*
Use ***CAB + The AI Decision Compass*** *together — one adds perspective, the other adds clarity.*

Problem: AI agrees too easily → ▦ Law 6 + Law 11

Problem: Generic responses → ▦ Law 2 + Law 14

Problem: Repetition or nonsense → ▦ Law 4 + Law 13

Problem: It feels too polite → ▦ Law 12 + Compass Framework

Problem: Ethical contradiction → ▦ Law 11 + Law 14

Problem: No usable format → ▦ Law 15 + Law 1

Problem: No evolution → ▦ Law 17 + Law 18

Problem: AI contradicts itself → ▦ Law 4 + Law 11 + Law 13

Problem: Too long / rambly output → ▦ Law 5 + Law 18

Problem: Output sounds robotic → ▦ Law 2 + Law 7

Problem: Summary instead of opinion → ▦ Law 9 + Law 17

Problem: Lacks depth or nuance → ▦ Law 12 + Law 3 + Law 14

Problem: Hallucinates / makes up facts → ▦ Law 3 + Law 13

Problem: Not adapting to feedback → ▦ Law 13 + Law 16

Problem: You're not sure what's missing → ▦ Law 12 + Law 9 + AI Compass Framework

Problem: Idea works once but not again → ▦ Law 17 + Law 4

Problem: Doesn't understand the task → ▦ Law 1 + Law 15

Problem: Sounds smart but feels wrong → ▦ Law 6 + Law 9

Problem: Too surface-level or safe → ▦ Law 14 + Compass Framework

Problem: You keep editing AI's work manually → ▦ Law 15 + Law 17

Problem: You want creative variation → ▦ Law 7 + Law 10

Problem: Needs input from multiple angles → ▦ Law 16 + CAB Framework

Problem: You're overwhelmed by too many directions → ▦ Law 1 + Law 15

Problem: You can't tell if the idea is actually good → ▦ Law 6 + Compass Framework

Problem: Prompt works in ChatGPT but not Claude/Gemini → ▦ Law 13 + 15

Epilogue – The Real Prompt Is You

Laws aren't commandments. They're launch codes.
These 19 were discovered through stress, iteration, and real-world collaboration.
But they're not finished. They evolve - just like you. **As we near AGI** and multimodal fluency, the future won't belong to those who prompt the most.
It'll belong to those who prompt with the most clarity, courage, and control.

Because every prompt is a mirror.
What we ask reveals what we assume.
What we clarify reveals what we value.
What we overlook? That's where the next evolution begins.

The genius of prompting isn't that it gets AI to think.
It's that it teaches you how to.

The Laws in this book weren't just written to train machines.
They were written to upgrade you.

To help you:

- Think like a strategist
- Question like a philosopher
- Guide like a leader
- And build like the future is already listening

Prompt wisely. Think bravely. Build the things the world doesn't know it needs.
Because the most powerful prompt isn't in this book.
It's the next one — the one only you can write.
The next move is yours.

✦ Acknowledgments

To my wife,
You are the quiet reason this book exists.
While the world talks about artificial intelligence, you reminded me—gently, daily - of what real intelligence looks like: love, patience, and perspective.
Thank you for making space for my late nights, endless drafts, and wild experiments. This book carries my name, but your fingerprint is on every page.

With deepest gratitude,
Nikolay

Quick Glossary Of Prompt Intelligence Terms

Prompt-as-a-Service (PaaS): Delivering prompts as reusable templates, tools, or assets.

Prompt Architecture: Structuring prompts modularly (intro, instruction, variables, constraints).

Recursive Prompting: When an AI self-improves or revises its own output based on re-prompting.

Prompt Portfolio: A library of curated, versioned, and tested prompts ready for reuse or sale.

Skeptic Simulation: A prompt technique where the AI simulates a critic to test assumptions.

Meta-Prompt: A prompt about prompting. Used to generate, explain, or refine other prompts.

Prompt Autopsy:
Analyzing a failed or underperforming prompt to uncover what went wrong and how to improve it.

Cognitive Clone:
An AI persona trained to mirror your thinking, writing, or problem-solving style— used to enhance creativity or scale yourself.

Delusional Prompt Layer:
A final prompt filter designed to detect hallucinations, exaggerations, or unsupported claims before publishing or using AI outputs.

Role Simulation Prompting:
Instructing the AI to think, write, or respond as a specific character, expert, persona, or psychological lens.

Prompt Chain Link:
Cross-referencing and combining multiple laws or prompt strategies for compounding results (e.g., Law 1 + Law 6 + Law 16 in sequence).

Prompt Audit:
Systematic review of saved prompts and outputs to evaluate performance, ethics, clarity, and potential for versioning.

Modular Prompting:
Designing prompts using swappable sections or building blocks that allow rapid adaptation for new audiences or tasks.

Prompt Debugging:
The process of identifying vague, biased, or poorly scoped instructions inside a prompt and correcting them.

Special Frameworks By Nikolay Gul

CAB (Cross-AI Brainstorming):
A collaborative prompting technique that merges outputs from multiple AI systems (e.g., GPT + Gemini + Claude) into a single, decision-optimized synthesis. Helps reduce bias and surface unseen insights. **Used in: Laws 10, 11, 13, 16; Bonus Chapter**

YIMBS – You Is Me, But Smarter. Now **"The AI Decision Compass":**
A personalized AI thinking framework where the AI adopts your exact personality, goals, and logic—but enhanced. Used for high-stakes decision-making, strategic writing, and thought delegation. **Used in: Laws 3, 7, 15, 17; Bonus Chapter**

PersonalIntel AI™:
An AI-powered digital profiling system that builds trust-based strategies by reverse-engineering how others perceive your public footprint—used for reputation building, positioning, and personal branding. Reverse-engineers how people (or AIs) perceive your digital presence. Used to build: Personal brand authority, Reputation trust signals, Influence positioning.
Used in: Law 9 (Mirror Self Protocol)

PROMPTING PIONEER RESOURCES

Books & Landmarks Papers:

- "The Alignment Problem" by Brian Christian
- "System Design for Large Language Models" – Anthropic Whitepaper
- "Designing with AI: Creative Co-Pilots" (MIT Press)

Frameworks, Tools & Links:

- Notion Prompt Library Template
- OpenAI Cookbook
- Anthropic's Claude Prompt Guide
- PromptHero.com – Multimodal prompt gallery

References & Credible Sources

Selected for credibility, thought leadership, and foundational impact

- Bostrom, N. (2014). *Superintelligence: Paths, Dangers, Strategies*. Oxford University Press.
- Marcus, G. & Davis, E. (2019). *Rebooting AI: Building Artificial Intelligence We Can Trust*. Pantheon.
- OpenAI (2023–2025). GPT-4 Technical Reports & Prompt Engineering Guides
- Google DeepMind (2024). *Prompting with Purpose: Cognitive Framing and LLMs*.
- Anthropic (2023). *Claude's Constitutional AI Whitepaper*.
- Stanford HAI. (2022–2025). *Human-Centered AI Policy and Design Reports

About the Author

Nikolay **Gul** is a marketing strategist, and prompt engineer, loke
clarity, creativity, and future-proof thinking. He is the author of *AI-
Driven Cybersecurity and High-Tech Marketing & Easy Book Self-
Publishing*, and dozens of frameworks used by professionals
worldwide.

Nikolay combines real-world industry experience with experimental
AI collaboration. His unique prompting frameworks — including *CAB
(Cross-AI Brainstorming)* and **The AI Decision Compass (*YIMBS*)**
and several more have been recognized for making AI more practical,
ethical, and empowering.
For inquiries or speaking: https://www.linkedin.com/in/webdesignerny

Selected Publications by the Author:
https://www.amazon.com/stores/Nikolay-Gul/author/B0DWCJZ8NV

**Gul, Nikolay. "*AI-Driven Cybersecurity and High-Tech Marketing: Practical,
Future-Proof Strategies for Experts and Beginners.*"**
→ ISBN: 979-8-218-61248-1 (Paperback)
→ ISBN: 979-8-9927440-0-2 (eBook)
→ LCCN: 2025902819

Gul, Nikolay. *"Easy Book Self-Publishing: A Step-by-Step Guide With AI
Assistance."*
|→ ISBN: 979-8-9927440-1-9 (Paperback)
→ ISBN: 979-8-9927440-2-6 (eBook)
→ LCCN: 2025906574

Gul, Nikolay. "The 19 Laws of AI Prompting Intelligence: *Master the Art of
Human-AI Thinking, Prompt Engineering, and Collaboration*"
→ ISBN: 979-8-9927440-4-0 (Paperback)
→ ISBN: 979-8-9927440-3-3 (eBook)
→ LCCN: 2025914398

Thank You to AI Tool Creators & Developers:

Special thanks to the developers, researchers, and open-source communities who built the tools that made this book possible — from AI-enhanced spellcheckers to frameworks like "Cross-AI Brainstorming" and "The AI Decision Compass."
This book was human-led, but AI-co-created — as a live demonstration of prompting intelligence in action.

Library Donation Note
I believe this book should be accessible to anyone helping shape the future of human–AI collaboration — especially students and educators.
If you're a professor, university librarian, or educational leader, I'd be happy to donate a printed copy for your institution's collection.
Just reach out to me directly via LinkedIn:
linkedin.com/in/webdesignerny

Connect with the author, Nikolay Gul,
on LinkedIn for more insights and updates:
https://www.linkedin.com/in/webdesignerny/

www.ingramcontent.com/pod-product-compliance
Lightning Source LLC
Chambersburg PA
CBHW071607210326
41597CB00019B/3445

9 798999 274040